Introduction

Mathematical Thinking at Grade 1

Grade 1

Marlene Kliman
Susan Jo Russell
Tracey Wright
Jan Mokros

Developed at TERC, Cambridge, Massachusetts

Dale Seymour Publications®
White Plains, New York

The *Investigations* curriculum was developed at TERC (formerly Technical Education Research Centers) in collaboration with Kent State University and the State University of New York at Buffalo. The work was supported in part by National Science Foundation Grant No. ESI-9050210. TERC is a nonprofit company working to improve mathematics and science education. TERC is located at 2067 Massachusetts Avenue, Cambridge, MA 02140.

This project was supported, in part,
by the
National Science Foundation
Opinions expressed are those of the authors
and not necessarily those of the Foundation

This book is published by Dale Seymour Publications®, an imprint of Addison Wesley Longman, Inc.

Dale Seymour Publications
10 Bank Street
White Plains, NY 10602

Managing Editor: Catherine Anderson
Series Editor: Beverly Cory
ESL Consultant: Nancy Sokol Green
Production/Manufacturing Director: Janet Yearian
Production/Manufacturing Manager: Karen Edmonds
Production/Manufacturing Coordinator: Shannon Miller
Design Manager: Jeff Kelly
Design: Don Taka
Composition: Shannon Miller, Andrea Reider
Illustrations: DJ Simison, Carl Yoshihara, Rachel Gage
Cover: Bay Graphics

DALE SEYMOUR PUBLICATIONS®

Order number DS43702
ISBN 1-57232- 466-X
3 4 5 6 7 8 9 10-ML-00 99 98

INVESTIGATIONS IN NUMBER, DATA, AND SPACE®

TERC

Principal Investigator   Susan Jo Russell

Co-Principal Investigator   Cornelia C. Tierney

Director of Research and Evaluation   Jan Mokros

Director of K–2 Curriculum   Karen Economopoulos

Curriculum Development

Karen Economopoulos
Marlene Kliman
Jan Mokros
Megan Murray
Susan Jo Russell
Tracey Wright

Evaluation and Assessment

Mary Berle-Carman
Jan Mokros
Andee Rubin

Teacher Support

Irene Baker
Megan Murray
Judy Storeygard
Tracey Wright

Technology Development

Michael T. Battista
Douglas H. Clements
Julie Sarama

Video Production

David A. Smith
Judy Storeygard

Administration and Production

Irene Baker
Amy Catlin

**Cooperating Classrooms
for This Unit**

Elizabeth A. Pedrini
Arlington Public Schools
Arlington, MA

Maryellen Bertrand
Mayra L. Cuevas
Barbara Reid
Boston Public Schools
Boston, MA

Nancy Frane
Joe Reilly
Malia Scott
Brookline Public Schools
Brookline, MA

Consultants and Advisors

Deborah Lowenberg Ball
Michael T. Battista
Marilyn Burns
Douglas H. Clements
Ann Grady

CONTENTS

WHERE TO START

The first-time user of *Mathematical Thinking at Grade 1* should read the following:

When you next teach this same unit, you can begin to read more of the background. Each time you present the unit, you will learn more about how your students understand the mathematical ideas.

Investigations in Number, Data, and Space® is a K–5 mathematics curriculum with four major goals:

- to offer students meaningful mathematical problems
- to emphasize depth in mathematical thinking rather than superficial exposure to a series of fragmented topics
- to communicate mathematics content and pedagogy to teachers
- to substantially expand the pool of mathematically literate students

The *Investigations* curriculum embodies an approach radically different from the traditional textbook-based curriculum. At each grade level, it consists of a set of separate units, each offering 2–8 weeks of work. These units of study are presented through investigations that involve students in the exploration of major mathematical ideas.

Approaching the mathematics content through investigations helps students develop flexibility and confidence in approaching problems, fluency in using mathematical skills and tools to solve problems, and proficiency in evaluating their solutions. Students also build a repertoire of ways to communicate about their mathematical thinking, while their enjoyment and appreciation of mathematics grow.

The investigations are carefully designed to invite all students into mathematics—girls and boys, members of diverse cultural, ethnic, and language groups, and students with different strengths and interests. Problem contexts often call on students to share experiences from their family, culture, or community. The curriculum eliminates barriers—such as work in isolation from peers, or emphasis on speed and memorization—that exclude some students from participating successfully in mathematics.

The following aspects of the curriculum ensure that all students are included in significant mathematics learning:

- Students spend time exploring problems in depth.
- They find more than one solution to many of the problems they work on.

- They invent their own strategies and approaches, rather than relying on memorized procedures.
- They choose from a variety of concrete materials and appropriate technology, including calculators, as a natural part of their everyday mathematical work.
- They express their mathematical thinking through drawing, writing, and talking.
- They work in a variety of groupings—as a whole class, individually, in pairs, and in small groups.
- They move around the classroom as they explore the mathematics in their environment and talk with their peers.

While reading and other language activities are typically given a great deal of time and emphasis in elementary classrooms, mathematics often does not get the time it needs. If students are to experience mathematics in depth, they must have enough time to become engaged in real mathematical problems. We believe that a minimum of 5 hours of mathematics classroom time a week—about an hour a day—is critical at the elementary level. The plan and pacing of the *Investigations* curriculum are based on that belief.

We explain more about the pedagogy and principles that underlie these investigations in Teacher Notes throughout the units. For correlations of the curriculum to the NCTM Standards and further help in using this research-based program for teaching mathematics, see the following books:

- *Implementing the* Investigations in Number, Data, and Space® *Curriculum*

- *Beyond Arithmetic: Changing Mathematics in the Elementary Classroom* by Jan Mokros, Susan Jo Russell, and Karen Economopoulos

This book is one of the curriculum units for *Investigations in Number, Data, and Space.* In addition to providing part of a complete mathematics curriculum for your students, this unit offers information to support your own professional development. You, the teacher, are the person who will make this curriculum come alive in the classroom; the book for each unit is your main support system.

Although the curriculum does not include student textbooks, reproducible sheets for student work are provided in the unit and are also available as Student Activity Booklets. Students work actively with objects and experiences in their own environment and with a variety of manipulative materials and technology, rather than with a book of instruction and problems. We strongly recommend use of the overhead projector as a way to present problems, to focus group discussion, and to help students share ideas and strategies.

Ultimately, every teacher will use these investigations in ways that make sense for his or her particular style, the particular group of students, and the constraints and supports of a particular school environment. Each unit offers information and guidance for a wide variety of situations, drawn from our collaborations with many teachers and students over many years. Our goal in this book is to help you, a professional educator, implement this curriculum in a way that will give all your students access to mathematical power.

Investigation Format

The opening two pages of each investigation help you get ready for the work that follows.

What Happens This gives a synopsis of each session or block of sessions.

Mathematical Emphasis This lists the most important ideas and processes students will encounter in this investigation.

What to Plan Ahead of Time These lists alert you to materials to gather, sheets to duplicate, transparencies to make, and anything else you need to do before starting.

INVESTIGATION 2

Exploring Numbers

What Happens

Session 1: The Game of Compare Students play the game Compare, in which they find the larger of two numbers. As students become ready for more challenge, either in this session or later in the investigation or unit, they play the game Double Compare, in which they find which of two totals is greater.

Sessions 2 and 3: Introducing Staircases and Choice Time Students are introduced to a new activity, building "staircases" from interlocking cubes. Then, they participate in Choice Time for the remainder of the sessions. The choices include building staircases and playing Compare or Double Compare.

Session 4: Seven Peas and Carrots Students solve a problem in which they have seven peas and carrots altogether. They determine one or more combinations of peas and carrots they could have to make up seven in all. This is the first of several How Many of Each? problems they will solve throughout the school year.

Sessions 5 and 6: Number Choices During Choice Time, students work on the following choices: solving a How Many of Each? problem involving nine peas and carrots, building staircases, and playing Compare or Double Compare. At the end of Session 6, they share ways that they recorded solutions to the How Many of Each? problem.

Routines Refer to the section About Classroom Routines (pp. 145–152) for suggestions on integrating into the school day regular practice of mathematical skills in counting, exploring data, and understanding time and changes.

Mathematical Emphasis

- Developing strategies for comparing two quantities up to about 20
- Using numbers to show how many
- Developing strategies for combining two single-digit numbers
- Finding combinations of numbers up to about 10
- Representing solutions to problems with pictures, numbers, and words
- Ordering a set of numbers up to about 20
- Counting up to 20 objects

INVESTIGATION 2

What to Plan Ahead of Time

Materials

- Chart paper or newsprint (18 by 24 inches): 15–20 sheets (available for teacher use)
- Blank letter-size paper (available for student use)
- Interlocking cubes: at least 30 per student
- Number Cards: 1 deck per pair, stored in resealable plastic bags or envelopes. If you do not have manufactured cards, make your own; see Other Preparation. (Sessions 1–3)
- Counters, such as buttons, bread tabs, or pennies: at least 30 per student (available for student use in all sessions). Be sure you have some counters in different colors.

Other Preparation

Duplicate the following student sheets and teaching resources, located at the end of this unit. If you have Student Activity Booklets, copy only the items marked with an asterisk.

For Session 1

Student Sheet 1, Compare (p. 157): 1 per student (homework)

Number Cards (pp. 174–177): 1 set per student (homework). If you do not have manufactured cards, you will also need a class set for each pair. Class sets will last longer if duplicated on card stock. Cut apart each set and store in a plastic resealable bag. A parent or other volunteer might help with the cutting.

For Sessions 2 and 3

Student Sheet 2, Double Compare (p. 158): 1 per student (homework)

Staircase Cards* (p. 160): 3–4 sheets, preferably on card stock. Cut apart each sheet to make two sets of 1–12 cards and two sets of 13–20 cards. (Use of the 13–20 set is optional.) Store each set in a plastic resealable bag.

Game Record Sheet* (p. 178): 1 per student (homework, optional)

For Sessions 5 and 6

Student Sheet 3, Bats and Balls (p. 159): 1 per student (homework). Consider modifying the number on this sheet for some students.

Sessions Within an investigation, the activities are organized by class session, a session being at least a one-hour math class. Sessions are numbered consecutively through an investigation. Often several sessions are grouped together, presenting a block of activities with a single major focus.

When you find a block of sessions presented together—for example, Sessions 1, 2, and 3—read through the entire block first to understand the overall flow and sequence of the activities. Make some preliminary decisions about how you will divide the activities into three sessions for your class, based on what you know about your students. You may need to modify your initial plans as you progress through the activities, and you may want to make notes in the margins of the pages as reminders for the next time you use the unit.

Be sure to read the Session Follow-Up section at the end of the session block to see what homework assignments and extensions are suggested as you make your initial plans.

While you may be used to a curriculum that tells you exactly what each class session should cover, we have found that the teacher is in a better position to make these decisions. Each unit is flexible and may be handled somewhat differently by every teacher. While we provide guidance for how many sessions a particular group of activities is likely to need, we want you to be active in determining an appropriate pace and the best transition points for your class. It is not unusual for a teacher to spend more or less time than is proposed for the activities.

Activities The activities include pair and small-group work, individual tasks, and whole-class discussions. In any case, students are seated together, talking and sharing ideas during all work times. Students most often work cooperatively, although each student may record work individually.

Choice Time In most units, some sessions are structured with activity choices. In these cases, students may work simultaneously on different activities focused on the same mathematical ideas. Students choose which activities they want to do, and they cycle through them. You will need to decide how to set up and introduce these activities and how to let students make their choices. Some

---

Session 1

The Game of Compare

What Happens

Students play the game Compare, in which they find the larger of two numbers. As students become ready for more challenge, either in this session or later in the investigation or unit, they play the game Double Compare, in which they find which of two totals is greater. Their work focuses on:

- playing a mathematical game with a partner
- comparing two numbers to find which is larger
- combining two quantities

Materials

- Number Cards, with wild cards removed (1 deck per pair, and 1 set per student for homework)
- Student Sheet 1. (1 per student, homework)
- Interlocking cubes (class set)
- Counters (such as buttons or bread tabs)

Activity

Compare

Compare is a number card game that students play in pairs. Introduce this game to the entire class by assembling students in a circle on the floor to watch a demonstration game, with either two student volunteers or you and a student as the two players. Have a deck of Number Cards with the wild cards removed.

Today we will play a game called Compare. At the beginning of the game, each player gets half the cards in the deck.

Demonstrate how to deal out the cards evenly between the two players. One student takes the deck and gives a card to his or her partner, and then takes a card. The student continues giving one card away and taking a card until all the cards have been distributed. (Each player should have 22 cards.)

You will turn your cards facedown. Then, you will both turn over your top card. If your number is larger than the other player's, you say "Me!" Let's see what happens when our volunteers do it.

After the volunteers turn over their first cards, ask the class which number is larger and how they know. Pause for a moment to ask students which card is larger and how they know.

Tuan turned over 7, and Tamika turned over 4. Which is larger? Who says "Me"?

Play two or three more turns, or until you think students understand the game.

---

teachers set up choices as stations around the room, while others post the list of available choices and allow students to collect their own materials and choose their own work space. You may need to experiment with a few different structures before finding a setup that works best for you.

Extensions These follow-up activities are opportunities for some or all students to explore a topic in greater depth or in a different context. They are not designed for "fast" students; mathematics is a multifaceted discipline, and different students will want to go further in different investigations. Look for and encourage the sparks of interest and enthusiasm you see in your students, and use the extensions to help them pursue these interests.

Excursions Some of the *Investigations* units include excursions—blocks of activities that could be omitted without harming the integrity of the unit. This is one way of dealing with the great depth and variety of elementary mathematics—much more than a class has time to explore in any one year. Excursions give you the flexibility to make different choices from year to year, doing the excursion in one unit this time, and next year trying another excursion.

Tips for the Linguistically Diverse Classroom At strategic points in each unit, you will find concrete suggestions for simple modifications of the teaching strategies to encourage the participation of all students. Many of these tips offer alternative ways to elicit critical thinking from students at varying levels of English proficiency, as well as from other students who find it difficult to verbalize their thinking.

The tips are supported by suggestions for specific vocabulary work to help ensure that all students can participate fully in the investigations. The Preview for the Linguistically Diverse Classroom (p. I-23) lists important words that are assumed as part of the working vocabulary of the unit. Second-language learners will need to become familiar with these words in order to understand the problems and activities they will be doing. These terms can be incorporated into students' second-language work before or during the unit. Activities that can be used to present the words are found in the appendix, Vocabulary Support for Second-Language Learners (p. 153). In addition, ideas for making connections to students' language and cultures, included on the Preview page, help the class explore the unit's concepts from a multicultural perspective.

Classroom Routines Activities in counting, exploring data, and understanding time and changes are suggested for routines in the grade 1 *Investigations* curriculum. Routines offer ongoing work with this important content as a regular part of the school day. Some routines provide more practice with content presented in the curriculum; others extend the curriculum; still others explore new content areas.

Plan to incorporate a few of the routine activities into a standard part of your daily schedule, such as morning meeting. When opportunities arise, you can also include routines as part of your work in other subject areas (for example, keeping a weather chart for science). Most routines are short and can be done whenever you have a spare 10-15 minutes, such as before lunch or recess or at the end of the day.

You will need to decide how often to present routines, what variations are appropriate for your class, and at what points in the day or week you will include them. A reminder about classroom routines is included on the first page of each investigation. Whatever routines you choose, your students will gain the most from these routines if they work with them regularly.

Materials

A complete list of the materials needed for teaching this unit is found on p. I-18. Some of these materials are available in kits for the *Investigations* curriculum. Individual items can also be purchased from school supply dealers.

Classroom Materials In an active mathematics classroom, certain basic materials should be available at all times: interlocking cubes, pencils, unlined paper, graph paper, calculators, and things to count with. Some activities in this curriculum require scissors and glue sticks or tape. Stick-on notes and large paper are also useful materials

throughout. So that students can independently get what they need at any time, they should know where these materials are kept, how they are stored, and how they are to be returned to the storage area. Many teachers have found that stopping 5 minutes before the end of each session so that students can finish their work and clean up is helpful in maintaining classroom materials. You'll find that establishing such routines at the beginning of the year is well worth the time and effort.

Technology Calculators are introduced to students in the first unit of the grade 1 sequence, *Mathematical Thinking at Grade 1*. By freely exploring and experimenting, students become familiar with this important mathematical tool.

Computer activities at grade 1 use a software program, called *Shapes,* that was developed especially for the *Investigations* curriculum. This program is introduced in the geometry unit, *Quilt Squares and Block Towns*. Using *Shapes,* students explore two-dimensional geometry while making pictures and designs with pattern block shapes and tangram pieces.

Although the software is linked to activities only in the geometry unit, we recommend that students use it throughout the year. Thus, you may want to introduce it when you introduce pattern blocks in *Mathematical Thinking at Grade 1*. How you use the computer activities depends on the number of computers you have available. Suggestions are offered in the geometry unit for how to organize different types of computer environments.

Children's Literature Each unit offers a list of suggested children's literature (p. I-18) that can be used to support the mathematical ideas in the unit. Sometimes an activity is based on a specific children's book, with suggestions for substitutions where practical. While such activities can be adapted and taught without the book, the literature offers a rich introduction and should be used whenever possible.

Student Sheets and Teaching Resources Student recording sheets and other teaching tools needed for both class and homework are provided as reproducible blackline masters at the end of each

unit. They are also available as Student Activity Booklets. These booklets contain all the sheets each student will need for individual work, freeing you from extensive copying (although you may need or want to copy the occasional teaching resource on transparency film or card stock, or make extra copies of a student sheet).

We think it's important that students find their own ways of organizing and recording their work. They need to learn how to explain their thinking with both drawings and written words, and how to organize their results so someone else can understand them. For this reason, we deliberately do not provide student sheets for every activity. Regardless of the form in which students do their work, we recommend that they keep a mathematics notebook or folder so that their work is always available for reference.

Homework In *Investigations,* homework is an extension of classroom work. Sometimes it offers review and practice of work done in class, sometimes preparation for upcoming activities, and sometimes numerical practice that revisits work in earlier units. Homework plays a role both in supporting students' learning and in helping inform families about the ways in which students in this curriculum work with mathematical ideas.

Depending on your school's homework policies and your own judgment, you may want to assign more homework than is suggested in the units. For this purpose you might use the practice pages, included as blackline masters at the end of this unit, to give students additional work with numbers.

For some homework assignments, you will want to adapt the activity to meet the needs of a variety of students in your class: those with special needs, those ready for more challenge, and second-language learners. You might change the numbers in a problem, make the activity more or less complex, or go through a sample activity with those who need extra help. You can modify any student sheet for either homework or class use. In particular, making numbers in a problem smaller or larger can make the same basic activity appropriate for a wider range of students.

Another issue to consider is how to handle the homework that students bring back to class—how to recognize the work they have done at home without spending too much time on it. Some teachers hold a short group discussion of different approaches to the assignment; others ask students to share and discuss their work with a neighbor, or post the homework around the room and give students time to tour it briefly. If you want to keep track of homework students bring in, be sure it ends up in a designated place.

Investigations at Home It is a good idea to make your policy on homework explicit to both students and their families when you begin teaching with *Investigations*. How frequently will you be assigning homework? When do you expect homework to be completed and brought back to school? What are your goals in assigning homework? How independent should families expect their children to be? What should the parent or guardian's role be? The more explicit you can be about your expectations, the better the homework experience will be for everyone.

Investigations at Home (a booklet available separately for each unit, to send home with students) gives you a way to communicate with families about the work students are doing in class. This booklet includes a brief description of every session, a list of the mathematics content emphasized in each investigation, and a discussion of each homework assignment to help families more effectively support their children. Whether or not you are using the *Investigations* at Home booklets, we expect you to make your own choices about homework assignments. Feel free to omit any and to add extra ones you think are appropriate.

Family Letter A letter that you can send home to students' families is included with the blackline masters for each unit. Families need to be informed about the mathematics work in your classroom; they should be encouraged to participate in and support their children's work. A reminder to send home the letter for each unit appears in one of the early investigations. These letters are also available separately in Spanish, Vietnamese, Cantonese, Hmong, and Cambodian.

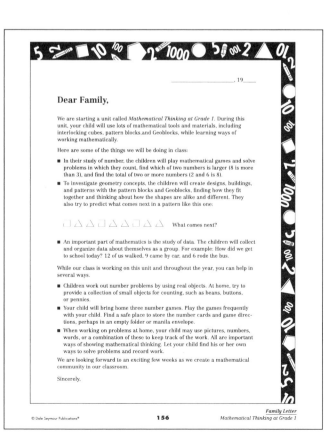

Help for You, the Teacher

Because we believe strongly that a new curriculum must help teachers think in new ways about mathematics and about their students' mathematical thinking processes, we have included a great deal of material to help you learn more about both.

About the Mathematics in This Unit This introductory section (p. I-19) summarizes the critical information about the mathematics you will be teaching. It describes the unit's central mathematical ideas and how students will encounter them through the unit's activities.

Teacher Notes These reference notes provide practical information about the mathematics you are teaching and about our experience with how students learn. Many of the notes were written in response to actual questions from teachers, or to discuss important things we saw happening in the field-test classrooms. Some teachers like to read them all before starting the unit, then review them as they come up in particular investigations.

Dialogue Boxes Sample dialogues demonstrate how students typically express their mathematical ideas, what issues and confusions arise in their thinking, and how some teachers have guided class discussions. These dialogues are based on the extensive classroom testing of this curriculum; many are word-for-word transcriptions of recorded class discussions. They are not always easy reading; sometimes it may take some effort to unravel what the students are trying to say. But this is the value of these dialogues; they offer good clues to how your students may develop and express their approaches and strategies, helping you prepare for your own class discussions.

Where to Start You may not have time to read everything the first time you use this unit. As a first-time user, you will likely focus on understanding the activities and working them out with your students. Read completely through each investigation before starting to present it. Also read those sections listed in the Contents under the heading Where to Start (p. vi).

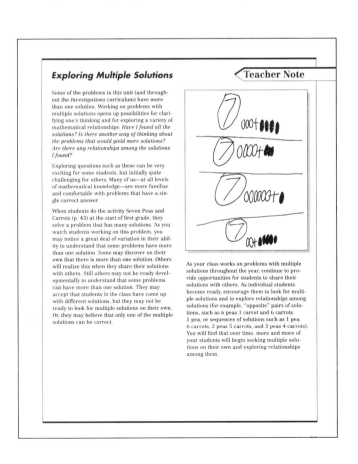

Exploring Multiple Solutions — Teacher Note

Some of the problems in this unit (and throughout the *Investigations* curriculum) have more than one solution. Working on problems with multiple solutions opens up possibilities for clarifying one's thinking and for exploring a variety of mathematical relationships: *Have I found all the solutions? Is there another way of thinking about the problems that would yield more solutions? Are there any relationships among the solutions I found?*

Exploring questions such as these can be very exciting for some students, but initially quite challenging for others. Many of us—at all levels of mathematical knowledge—are more familiar and comfortable with problems that have a single correct answer.

When students do the activity Seven Peas and Carrots (p. 43) at the start of first grade, they solve a problem that has many solutions. As you watch students working on this problem, you may notice a great deal of variation in their ability to understand that some problems have more than one solution. Some may discover on their own that there is more than one solution. Others will realize this when they share their solutions with others. Still others may not be ready developmentally to understand that some problems can have more than one solution. They may accept that students in the class have come up with different solutions, but they may not be ready to look for multiple solutions on their own. Or, they may believe that only one of the multiple solutions can be correct.

As your class works on problems with multiple solutions throughout the year, continue to provide opportunities for students to share their solutions with others. As individual students become ready, encourage them to look for multiple solutions and to explore relationships among solutions (for example, "opposite" pairs of solutions, such as 6 peas 1 carrot and 6 carrots 1 pea, or sequences of solutions such as 1 pea 6 carrots, 2 peas 5 carrots, and 3 peas 4 carrots). You will find that over time, more and more of your students will begin seeking multiple solutions on their own and exploring relationships among them.

DIALOGUE BOX

What Comes Next?

Students in this class are learning how to play What Comes Next? with their teacher (p. 62). Each student has a paper cup with one green cube, one white cube, and one orange cube. The class has already worked with a-b-a-b-a-b patterns, and this time the teacher has hidden an a-b-c-a-b-c pattern in the tube: white-green-orange-white-green-orange. The first four cubes are showing.

So we have white-green-orange-white so far. Hold up the cube you think comes next. *[Pause.]* **Yanni, why did you pick green?**

Yanni: Because a pattern keeps repeating.

Tamika: It's got another pattern. White, green, orange and it keeps on repeating itself over and over.

Anyone have something to add? Andre, you picked green, too. Why?

Andre: Because the green comes after the white.

Libby: Because the pattern goes white, green, orange, white, green, orange.

OK. Here's a new pattern. I'm not going to start until all the cubes are back in the cups.

The teacher shows two cubes, green followed by orange. There are a lot of different guesses of what might come next; all three colors are named as possibilities. The teacher reveals that the third cube is orange; the cubes showing now are green-orange-orange. Most students predict a green cube next, although Fernando shows a white one.

Who wants to tell us a reason for the color they're holding up?

Claire: I think green goes next. The pattern goes green, orange, orange, green, orange, orange.

Kaneisha: I picked green because I knew it was green at the top and then orange orange, so next would have to be green.

Why did you choose white, Fernando?

Fernando: Well, it's probably green, like Kaneisha said, but it could be white. I was thinking green-orange-orange and then two whites and then it would start again.

So Fernando is saying it's possible that we may not have seen the whole pattern yet. It could go green-orange-orange-white, then green-orange-orange-white-white again. So, let's see one more cube, and then see what everyone thinks.

Most of these students seem to be able to think about a pattern made from a three-element unit. Fernando is thinking of even more possibilities. He seems to have the idea that the unit of a pattern can be made out of even more than three cubes, as he suggests a possible five-element unit. The teacher knows that not all students are following Fernando's explanation, but restates his idea for those students who might begin to consider it.

The *Investigations* curriculum incorporates the use of two forms of technology in the classroom: calculators and computers. Calculators are assumed to be standard classroom materials, available for student use in any unit. Computers are explicitly linked to one or more units at each grade level; they are used with the unit on 2-D geometry unit at each grade, as well as with some of the units on measuring, data, and changes.

Using Calculators

In this curriculum, calculators are considered tools for doing mathematics, similar to pattern blocks or interlocking cubes. Just as with other tools, students must learn both *how* to use calculators correctly and *when* they are appropriate to use. This knowledge is crucial for daily life, as calculators are now a standard way of handling numerical operations, both at work and at home.

Using a calculator correctly is not a simple task; it depends on a good knowledge of the four operations and of the number system, so that students can select suitable calculations and also determine what a reasonable result would be. These skills are the basis of any work with numbers, whether or not a calculator is involved.

Unfortunately, calculators are often seen as tools to check computations with, as if other methods are somehow more fallible. Students need to understand that any computational method can be used to check any other; it's just as easy to make a mistake on the calculator as it is to make a mistake on paper or with mental arithmetic. Throughout this curriculum, we encourage students to solve computation problems in more than one way in order to double-check their accuracy. We present mental arithmetic, paper-and-pencil computation, and calculators as three possible approaches.

In this curriculum we also recognize that, despite their importance, calculators are not always appropriate in mathematics instruction. Like any tools, calculators are useful for some tasks, but not for others. You will need to make decisions about when to allow students access to calculators and when to ask that they solve problems without them, so that they can concentrate on other tools and skills. At times when calculators are or are not appropriate for a particular activity, we make specific recommendations. Help your students develop their own sense of which problems they can tackle with their own reasoning and which ones might be better solved with a combination of their own reasoning and the calculator.

Managing calculators in your classroom so that they are a tool, and not a distraction, requires some planning. When calculators are first introduced, students often want to use them for everything, even problems that can be solved quite simply by other methods. However, once the novelty wears off, students are just as interested in developing their own strategies, especially when these strategies are emphasized and valued in the classroom. Over time, students will come to recognize the ease and value of solving problems mentally, with paper and pencil, or with manipulatives, while also understanding the power of the calculator to facilitate work with larger numbers.

Experience shows that if calculators are available only occasionally, students become excited and distracted when they are permitted to use them. They focus on the tool rather than on the mathematics. In order to learn when calculators are appropriate and when they are not, students must have easy access to them and use them routinely in their work.

If you have a calculator for each student, and if you think your students can accept the responsibility, you might allow them to keep their calculators with the rest of their individual materials, at least for the first few weeks of school. Alternatively, you might store them in boxes on a shelf, number each calculator, and assign a corresponding number to each student. This system can give students a sense of ownership while also helping you keep track of the calculators.

Using Computers

Students can use computers to approach and visualize mathematical situations in new ways. The computer allows students to construct and manipulate geometric shapes, see objects move according to rules they specify, and turn, flip, and repeat a pattern.

This curriculum calls for computers in units where they are a particularly effective tool for learning mathematics content. One unit on 2-D geometry at each of the grades 3–5 includes a core of activities that rely on access to computers, either in the classroom or in a lab. Other units on geometry, measurement, data, and changes include computer activities, but can be taught without them. In these units, however, students' experience is greatly enhanced by computer use.

The following list outlines the recommended use of computers in this curriculum:

Grade 1
Unit: *Survey Questions and Secret Rules*
  (Collecting and Sorting Data)
Software: Tabletop, Jr.
Source: Broderbund

Unit: *Quilt Squares and Block Towns*
  (2-D and 3-D Geometry)
Software: *Shapes*
Source: provided with the unit

Grade 2
Unit: *Mathematical Thinking at Grade 2*
  (Introduction)
Software: *Shapes*
Source: provided with the unit

Unit: *Shapes, Halves, and Symmetry*
  (Geometry and Fractions)
Software: *Shapes*
Source: provided with the unit

Unit: *How Long? How Far?* (Measuring)
Software: *Geo-Logo*
Source: provided with the unit

Grade 3
Unit: *Flips, Turns, and Area* (2-D Geometry)
Software: *Tumbling Tetrominoes*
Source: provided with the unit

Unit: *Turtle Paths* (2-D Geometry)
Software: *Geo-Logo*
Source: provided with the unit

Grade 4
Unit: *Sunken Ships and Grid Patterns*
  (2-D Geometry)
Software: *Geo-Logo*
Source: provided with the unit

Grade 5
Unit: *Picturing Polygons* (2-D Geometry)
Software: *Geo-Logo*
Source: provided with the unit

Unit: *Patterns of Change* (Tables and Graphs)
Software: *Trips*
Source: provided with the unit

Unit: *Data: Kids, Cats, and Ads* (Statistics)
Software: Tabletop, Sr.
Source: Broderbund

The software provided with the *Investigations* units uses the power of the computer to help students explore mathematical ideas and relationships that cannot be explored in the same way with physical materials. With the *Shapes* (grades 1–2) and *Tumbling Tetrominoes* (grade 3) software, students explore symmetry, pattern, rotation and reflection, area, and characteristics of 2-D shapes. With the *Geo-Logo* software (grades 3–5), students investigate rotations and reflections, coordinate geometry, the properties of 2-D shapes, and angles. The *Trips* software (grade 5) is a mathematical exploration of motion in which students run experiments and interpret data presented in graphs and tables.

We suggest that students work in pairs on the computer; this not only maximizes computer resources but also encourages students to consult, monitor, and teach one another. Generally, more than two students at one computer find it difficult to share. Managing access to computers is an issue for every classroom. The curriculum gives you explicit support for setting up a system. The units are structured on the assumption that you have enough computers for half your students to work on the machines in pairs at one time. If you do not have access to that many computers, suggestions are made for structuring class time to use the unit with five to eight computers, or even with fewer than five.

Assessment plays a critical role in teaching and learning, and it is an integral part of the *Investigations* curriculum. For a teacher using these units, assessment is an ongoing process. You observe students' discussions and explanations of their strategies on a daily basis and examine their work as it evolves. While students are busy recording and representing their work, working on projects, sharing with partners, and playing mathematical games, you have many opportunities to observe their mathematical thinking. What you learn through observation guides your decisions about how to proceed. In any of the units, you will repeatedly consider questions like these:

■ Do students come up with their own strategies for solving problems, or do they expect others to tell them what to do? What do their strategies reveal about their mathematical understanding?

■ Do students understand that there are different strategies for solving problems? Do they articulate their strategies and try to understand other students' strategies?

■ How effectively do students use materials as tools to help with their mathematical work?

■ Do students have effective ideas for keeping track of and recording their work? Does keeping track of and recording their work seem difficult for them?

You will need to develop a comfortable and efficient system for recording and keeping track of your observations. Some teachers keep a clipboard handy and jot notes on a class list or on adhesive labels that are later transferred to student files. Others keep loose-leaf notebooks with a page for each student and make weekly notes about what they have observed in class.

Assessment Tools in the Unit

With the activities in each unit, you will find questions to guide your thinking while observing the students at work. You will also find two built-in assessment tools: Teacher Checkpoints and embedded Assessment activities.

Teacher Checkpoints The designated Teacher Checkpoints in each unit offer a time to "check in" with individual students, watch them at work, and ask questions that illuminate how they are thinking.

At first it may be hard to know what to look for, hard to know what kinds of questions to ask. Students may be reluctant to talk; they may not be accustomed to having the teacher ask them about their work, or they may not know how to explain their thinking. Two important ingredients of this process are asking students open-ended questions about their work and showing genuine interest in how they are approaching the task. When students see that you are interested in their thinking and are counting on them to come up with their own ways of solving problems, they may surprise you with the depth of their understanding.

Teacher Checkpoints also give you the chance to pause in the teaching sequence and reflect on how your class is doing overall. Think about whether you need to adjust your pacing: Are most students fluent with strategies for solving a particular kind of problem? Are they just starting to formulate good strategies? Or are they still struggling with how to start? Depending on what you see as the students work, you may want to spend more time on similar problems, change some of the problems to use smaller numbers, move quickly to more challenging material, modify subsequent activities for some students, work on particular ideas with a small group, or pair students who have good strategies with those who are having more difficulty.

Embedded Assessment Activities Assessment activities embedded in each unit will help you examine specific pieces of student work, figure out what it means, and provide feedback. From the students' point of view, these assessment activities are no different from any others. Each is a learning experience in and of itself, as well as an opportunity for you to gather evidence about students' mathematical understanding.

The embedded assessment activities sometimes involve writing and reflecting; at other times, a discussion or brief interaction between student and teacher; and in still other instances, the creation and explanation of a product. In most cases, the assessments require that students *show* what they did, *write* or *talk* about it, or do both. Having to explain how they worked through a problem helps students be more focused and clear in their mathematical thinking. It also helps them realize that doing mathematics is a process that may involve tentative starts, revising one's approach, taking different paths, and working through ideas.

Teachers often find the hardest part of assessment to be interpreting their students' work. We provide guidelines to help with that interpretation. If you have used a process approach to teaching writing, the assessment in *Investigations* will seem familiar. For many of the assessment activities, a Teacher Note provides examples of student work and a commentary on what it indicates about student thinking.

Documentation of Student Growth

To form an overall picture of mathematical progress, it is important to document each student's work in journals, notebooks, or portfolios. The choice is largely a matter of personal preference; some teachers have students keep a notebook or folder for each unit, while others prefer one mathematics notebook, or a portfolio of selected work for the entire year. The final activity in each *Investigations* unit, called Choosing Student Work to Save, helps you and the students select representative samples for a record of their work.

This kind of regular documentation helps you synthesize information about each student as a mathematical learner. From different pieces of evidence, you can put together the big picture. This synthesis will be invaluable in thinking about where to go next with a particular child, deciding where more work is needed, or explaining to parents (or other teachers) how a child is doing.

If you use portfolios, you need to collect a good balance of work, yet avoid being swamped with an overwhelming amount of paper. Following are some tips for effective portfolios:

- Collect a representative sample of work, including some pieces that students themselves select for inclusion in the portfolio. There should be just a few pieces for each unit, showing different kinds of work—some assignments that involve writing, as well as some that do not.
- If students do not date their work, do so yourself so that you can reconstruct the order in which pieces were done.
- Include your reflections on the work. When you are looking back over the whole year, such comments are reminders of what seemed especially interesting about a particular piece; they can also be helpful to other teachers and to parents. Older students should be encouraged to write their own reflections about their work.

Assessment Overview

There are two places to turn for a preview of the assessment opportunities in each *Investigations* unit. The Assessment Resources column in the unit Overview Chart (pp. I-13–I-17) identifies the Teacher Checkpoints and Assessment activities embedded in each investigation, guidelines for observing the students that appear within classroom activities, and any Teacher Notes and Dialogue Boxes that explain what to look for and what types of student responses you might expect to see in your classroom. Additionally, the section About the Assessment in This Unit (p. I-21) gives you a detailed list of questions for each investigation, keyed to the mathematical emphases, to help you observe student growth.

Depending on your situation, you may want to provide additional assessment opportunities. Most of the investigations lend themselves to more frequent assessment, simply by having students do more writing and recording while they are working.

Mathematical Thinking at Grade 1

Content of This Unit This unit introduces first graders to some of the mathematical materials and processes they will be using this year as they explore counting, comparing, and combining in each of the three areas of the *Investigations* curriculum: number, data, and space.

Students use such mathematical tools and materials as interlocking cubes, pattern blocks, and Geoblocks as they count, combine numbers, play mathematical games, solve problems, investigate patterns, represent the results of surveys they take, and create their own designs, buildings, and constructions. They are also engaged in critical mathematical processes such as sharing and explaining their strategies; using pictures, numbers, and words to show their work; and working with peers.

This unit is designed to help you get to know and to assess your students' mathematical understanding and to help you establish a mathematical community in your classroom. It includes suggestions for organizing the classroom environment and for establishing classroom routines.

Connections with Other Units If you are doing the full-year *Investigations* curriculum in the suggested sequence for grade 1, this is the first of six units. It has connections with every other unit in the first grade sequence, both in its content and in its emphasis on ways of thinking and working. For example, in this unit students begin developing strategies for combining two quantities. They will continue to expand and refine their strategies in the two number units, *Building Number Sense* and *Number Games and Story Problems*. Also in this first unit, they explore geometric materials they will use again in *Quilt Squares and Block Towns* to compare and describe two- and three-dimensional shapes, and they begin to collect and represent survey data, which is a focus of the unit *Survey Questions and Secret Rules*.

Investigations Curriculum ▪ Suggested Grade 1 Sequence

▶ *Mathematical Thinking at Grade 1* (Introduction)

Building Number Sense (The Number System)

Survey Questions and Secret Rules (Collecting and Sorting Data)

Quilt Squares and Block Towns (2-D and 3-D Geometry)

Number Games and Story Problems (Addition and Subtraction)

Bigger, Taller, Heavier, Smaller (Measuring)

Investigation 1 ▪ Exploring Materials

Class Sessions	Activities	Pacing
Session 1 (p. 4) GETTING STARTED WITH MATERIALS	Building Things What Did You Notice?	minimum 1 hr
Sessions 2, 3, and 4 (p. 13) EXPLORING MATERIALS	Exploring Calculators Buildings, Patterns, and Calculators Teacher Checkpoint: Counting 20 Homework: Family Connection	minimum 3 hr
Classroom Routines		

Mathematical Emphasis

- Exploring mathematical materials and tools, such as pattern blocks, interlocking cubes, Geoblocks, and calculators

- Comparing and finding relationships among geometric shapes

Assessment Resources

Observing the Students (p. 6)

Talking About Shapes (Teacher Note, p. 11)

Teacher Checkpoint: Counting 20 (p. 17)

Observing Students Counting (Teacher Note, p. 19)

Mine Is a Sailboat (Dialogue Box, p. 23)

Materials

Interlocking cubes

Pattern blocks

Geoblocks

Scoops

Tubs or shoe boxes

Calculators

Family letter

Investigation 2 ■ Exploring Numbers

Class Sessions	Activities	Pacing
Session 1 (p. 26) THE GAME OF COMPARE	Compare Double Compare Homework: Compare	minimum 1 hr
Sessions 2 and 3 (p. 34) INTRODUCING STAIRCASES AND CHOICE TIME	Introducing Staircases Introducing Choice Time Homework: Compare or Double Compare Extension: Triple Compare	minimum 2 hr
Session 4 (p. 43) SEVEN PEAS AND CARROTS	Teacher Checkpoint: Seven Peas and Carrots Sharing Solutions Extension: Finding All the Solutions Extension: Peas, Carrots, and Blueberries Extension: Ways to Make Seven	minimum 1 hr
Sessions 5 and 6 (p. 54) NUMBER CHOICES	Choice Time Sharing Recording Methods Homework: Balls and Bats	minimum 2 hr

Classroom Routines

Mathematical Emphasis

- Developing strategies for comparing two quantities up to about 20

- Using numbers to show how many

- Developing strategies for combining two single-digit numbers

- Finding combinations of numbers up to 10

- Representing solutions to mathematics problems with pictures, numbers, and words

Assessment Resources

Observing the Students (pp. 27, 37, 55)

Double Compare: Strategies for Combining and Comparing (Teacher Note, p. 30)

First Graders: A Wide Range of Understanding (Teacher Note, p. 33)

Teacher Checkpoint: Seven Peas and Carrots (p. 43)

How Students Approach Seven Peas and Carrots (Teacher Note, p. 50)

Materials

Number Cards

Interlocking cubes

Counters

Staircase Cards

Student Sheets 1–3

Teaching resource sheets

Investigation 3 ■ Patterns

Class Sessions	Activities	Pacing
Session 1 (p. 62) WHAT COMES NEXT?	Cube Patterns Make Your Own Cube Pattern	minimum 1 hr
Session 2 (p. 67) CLAPPING PATTERNS	Clapping Patterns What Comes Next? Homework: Compare or Double Compare	minimum 1 hr
Sessions 3 and 4 (p. 69) FINDING AND MAKING PATTERNS	Same or Different? Choice Time Patterns Around Us Where Does This Pattern Fit? Homework: Patterns from Home	minimum 2 hr
Sessions 5 and 6 (p. 77) WHAT IS A PATTERN?	Starting the Pattern Exhibit Choice Time What's a Pattern?	minimum 2 hr

Classroom Routines

Mathematical Emphasis

- Describing pattern sequences
- Predicting what comes next in a pattern sequence
- Constructing patterns from a variety of materials

Assessment Resources

Seeing Patterns (Teacher Note, p. 65)

What Comes Next? (Dialogue Box, p. 66)

Observing the Students (pp. 72, 80)

First Graders' Cube Patterns (Teacher Note, p. 75)

Materials

Interlocking cubes

Pattern blocks

Paper cups

Scissors

Crayons or markers

Collage materials for making patterns

Geoblocks

Student Sheet 4

Teaching resource sheets

Investigation 4 ■ Counting and Combining

Class Sessions	Activities	Pacing
Session 1 (p. 84) COLLECT 15 TOGETHER	Collect 15 Together Homework: Collect 15 Together	minimum 1 hr
Sessions 2 and 3 (p. 88) COUNTING AND COMBINING	Choice Time Homework: Math Games Extension: Ways to Make 15	minimum 2 hr
Session 4 (p. 92) ELEVEN FRUITS	Assessment: 11 Fruits: How Many of Each? Sharing Solution Methods Homework: Dinosaurs and Tigers	minimum 1 hr
Session 5 (p. 100) MAKING PREDICTIONS ABOUT A STORY	A Story with a Growing Pattern What Could Happen Next?	minimum 1 hr
Session 6 (p. 106) HOW MANY IN ALL?	Teacher Checkpoint: How Many in All? Extension: Exploring Other Totals in the Story	minimum 1 hr

Classroom Routines

Mathematical Emphasis

- Counting and keeping track of a set of objects

- Extending and deepening understanding of comparing two quantities

- Using counting, patterns, and other strategies to help solve problems

- Extending and deepening understanding of number combinations

- Extending and deepening strategies for combining two quantities

- Representing solutions to mathematics problems with pictures, numbers, and words

- Making and explaining predictions

Assessment Resources

Observing the Students
(pp. 85, 90, 102)

Assessment:
11 Fruits: How Many of Each?
(p. 92)

Solution Strategies for 11 Fruits
(Dialogue Box, p. 98)

Teacher Checkpoint: How Many in All? (p. 106)

How Many in All? (Dialogue Box, p. 111)

Materials

Dot cubes

Pattern blocks

Geoblocks

Interlocking cubes

Number Cards

Counters

Rooster's Off to See the World or similar book

Student Sheets 5–6

Teaching resource sheets

Investigation 5 ▪ Data About Our Class

Class Sessions	Activities	Pacing
Session 1 (p. 116) KID PINS	How Many Are Here Today? Making Kid Pins	minimum 1 hr
Session 2 (p. 119) ATTENDANCE AND OTHER SURVEYS	Quick Surveys Attendance Survey Homework: Math Games Extension: More Quick Surveys	minimum 1 hr
Sessions 3 and 4 (p. 127) INVENTING SURVEY REPRESENTATIONS	Getting Attendance Data Sorting Ourselves Creating Representations of Data Barefoot Survey: Recording and Representing Results Homework: How Many of Each? Extension: Sharing Results of Data Surveys	minimum 2 hr
Sessions 5 and 6 (p. 135) HOW WE GOT TO SCHOOL TODAY	Ways to Get to School Collecting the Data Teacher Checkpoint: Representations of How We Got to School Discussing the Representations Choosing Student Work to Save	minimum 2 hr

Classroom Routines

Mathematical Emphasis

- Inventing representations that show what a survey was about

- Categorizing, data in ways that communicate clearly to others

- Representing the sizes of different groups

- Counting, combining, and comparing the sizes of different groups

- Making sense of survey results and presenting them to others

Assessment Resources

Data Categories (Teacher Note, p. 125)

Making Sense of Kid Pin Data (Dialogue Box, p. 126)

Observing the Students (p. 130)

Surveys and Representations (Dialogue Box, p. 134)

Teacher Checkpoint: Representations of How We Got to School (p. 137)

Student Representations of Getting to School (Teacher Note, p. 140)

Sharing Representations (Dialogue Box, p. 142)

What Stands for What? (Dialogue Box, p. 144)

Materials

Kid Pins (made of tongue depressors, spring-clip clothespins, markers, glue)

Survey boards (cardboard or foam core)

Interlocking cubes

Class lists

Drawing paper

Colored pencils or markers

Construction paper, stickers, buttons, glue

Student Sheets 7–9

Teaching resource sheets

Following are the basic materials needed for the activities in this unit. Many of the items can be purchased from the publisher, either individually or in the Teacher Resource Package and the Student Materials Kit for grade 1. Detailed information is available on the *Investigations* order form. To obtain this form, call toll-free 1-800-872-1100 and ask for a Dale Seymour customer service representative.

Interlocking cubes (cubes that connect on all sides): a class supply with at least 30 per student

Pattern blocks: 1 bucket per 6–8 students

Geoblocks: 1–2 sets per classroom

Calculators: at least 1 per pair

Counters, such as buttons, bread tabs, or pennies: at least 30 per student

Primary Number Cards (referred to as Number Cards in the unit): 1 deck per pair (manufactured, or use blackline masters to make your own)

Dot cubes: 1 per pair

Paper cups: 1 per student

Rooster's Off to See the World by Eric Carle (optional)

This Is the Way We Go to School by Edith Baer (optional)

Collage or construction materials for making patterns, such as paper or fabric scraps, buttons, beads, sequins, pasta shapes

Heavy cardboard or foam core: 2 large sheets (about 12 by 24 inches)

Tongue depressors: 1 per student

Spring-clip wooden clothespins: 1 per student

Glue (fast-drying)

Crayons, markers, or colored pencils

Scissors

Chart paper or newsprint (18 by 24 inches)

Large paper (11 by 17 inches): 3 sheets per student

Construction paper, colored dot stickers (optional)

Stick-on notes in a variety of sizes

Tubs or shoe boxes (for storing materials), and small containers such as plastic cups or yogurt cartons (for scooping materials)

Resealable plastic bags (for storing cards)

The following materials are provided at the end of this unit as blackline masters. A Student Activity Booklet containing all student sheets and teaching resources needed for individual work is available.

Family Letter (p. 156)

Student Sheets 1–9 (p. 157)

Teaching Resources:
    Staircase Cards (p. 160)
    Cube Pattern Strips (p. 162)
    Pattern Block Cutouts (p. 163)
    Number Cards (p. 174)
    Game Record Sheet (p. 178)
Practice Pages (p. 179)

Related Children's Literature

Exploring Materials
Stevenson, Robert Louis. *Block City*. New York: E. P. Dutton, 1988.

Numbers
Crews, Donald. *Ten Black Dots*. New York: Greenwillow Books, 1986.

Falwell, Cathryn. *Feast for Ten*. New York: Clarion Books, 1993.

Micklethwait, Lucy. *I Spy Two Eyes: Numbers in Art*. New York: Greenwillow Books, 1992.

Onyefulu, Ifeoma. *Emeka's Gift: An African Counting Story*. New York: Cobblehill Books/Dutton, 1995.

Pattern
Grossman, Virginia and Sylvia Long. *Ten Little Rabbits*. San Francisco: Chronicle Books, 1991.

Pluckrose, Henry. *Math Counts: Pattern*. Chicago: Childrens Press, 1995.

Stories with Growing Patterns
Carle Eric. *Rooster's Off to See the World*. New York: Picture Book Studio, 1987. Originally published as *The Rooster Who Set Out to See the World*, New York: Franklin Watts, 1972.

Carle, Eric. *The Very Hungry Caterpillar*. New York: Philomel Books, Putnam, 1981.

Lewis, Paul Owen. *P. Bear's New Year's Party*. Hillsboro, OR: Beyond Words Publishing, 1989.

Data
Baer, Edith. *This Is the Way We Go to School: A Book About Children Around the World*. New York: Scholastic, 1990.

Mathematical Thinking at Grade 1 is an introduction to mathematical content, materials, processes, and ways of working. Through the work in this unit, students

- solve mathematical problems in ways that make sense to them
- explore materials and use them to build models of mathematical situations
- talk, draw, and write about their work
- work with peers
- rely on their own thinking, and learn from the thinking of others

Students count and compare quantities in each of the three areas of the *Investigations* curriculum: number, space (geometry), and data. For their work with number, they play games and solve problems that involve counting, determine which of two numbers is larger, find the total of two or more numbers, and find numbers that sum to a given total such as 7.

In the geometry activities, students create designs, buildings, and patterns with two- and three-dimensional shapes, explore ways that shapes fit together, and find ways that shapes are alike and different. They also compare patterns and play games that involve predicting what comes next in a pattern.

Finally, in their work with data, students collect, count, organize, and represent data about themselves as a group, taking simple surveys to find, for example, who is wearing shoelaces and who isn't. As they talk about the results of their surveys, they compare groups to determine which is larger.

As students work on these activities, they become familiar with a variety of mathematical tools and materials, including interlocking cubes, pattern blocks, and Geoblocks. They use these materials to build things, to help them solve problems, to create patterns, and to represent the results of surveys they take.

Throughout this unit, students develop and use good number sense to combine one-digit numbers and to count and compare numbers up to at least 20. Just as common sense grows from experience with the world and how it works, number sense grows from experience with how numbers work.

While almost all students entering first grade know the oral counting sequence, they vary tremendously in their ability to accurately count out a set of about 20 objects and in their sense of the size of quantities. For instance, they may be able to say the counting sequence up to 20, but they may not understand that when they count one more, they are referring to a quantity that has one more.

Students are encouraged to build upon what they do know in ways that are challenging but not threatening. For example, in one game, Double Compare, students find the total of two one-digit numbers. Some students use objects to help them combine the numbers, some develop strategies that involve counting orally or on their fingers, and some begin to remember combinations or find ways to use a known combination to find others.

In the activity Seven Peas and Carrots, students are told they have seven things in all, some of which are peas and some of which are carrots. They find a combination of peas and carrots that would make seven in all. Some students solve the problem by counting out objects, others by drawing pictures, and others by using knowledge of number combinations. Some students find just one solution, others find several, and still others challenge themselves to find them all.

Mathematical Thinking at Grade 1 not only involves students with some central mathematical concepts, but also introduces them to a particular way of approaching mathematics. Throughout the unit students are encouraged to share their strategies, work cooperatively, use materials, and communicate both orally and on paper about how they are solving problems. These approaches may be quite difficult for some students; even the process of taking out, using, and putting away materials may be unfamiliar. Certainly drawing pictures or writing to describe mathematical thinking will be hard for some students.

This unit is a time to focus on the development of these processes; to spend time establishing routines and expectations; to communicate to students your own interest in and respect for their mathematical ideas; to assure students that you want to know about their *thinking,* not just their answers; and to insist that students work hard to solve problems in ways that make sense to them. As the unit unfolds, a mathematical community begins to take shape—a community that you and your students are together responsible for creating and maintaining.

Mathematical Emphasis At the beginning of each investigation, the Mathematical Emphasis section tells you what is most important for students to learn about during that investigation. Many of these understandings and processes are difficult and complex. Students gradually learn more and more about each idea over many years of schooling. Individual students will begin and end the unit with different levels of knowledge and skill, but all will learn more about mathematical thinking in the areas of number, data, and space (geometry), and will become more comfortable talking about and showing their ways of solving mathematical problems.

Throughout the *Investigations* curriculum, there are many opportunities for ongoing daily assessment as you observe, listen to, and interact with students at work. In this unit you will find four Teacher Checkpoints:

Investigation 1, Sessions 2–4:
Counting 20 (p. 17)

Investigation 2, Session 4:
Seven Peas and Carrots (p. 43)

Investigation 4, Session 6:
How Many in All? (p. 106)

Investigation 5, Sessions 5–6:
Representations of How We Got to School
(p. 137)

This unit also has one embedded assessment activity:

Investigation 4, Session 4:
11 Fruits: How Many of Each? (p. 92)

In addition, you can use almost any activity in this unit to assess your students' needs and strengths. Listed below are questions to help you focus your observations in each investigation. You may want to keep track of your observations for each student to help you plan your curriculum and monitor students' growth. Suggestions for documenting student growth can be found in the section About Assessment (p. I-10).

Investigation 1: Exploring Materials

■ Are students already familiar with any of the materials? How do students explore the materials? Do they initiate their own ideas, observe others, or follow given prompts or suggestions? Can they follow an idea through to completion, or do they begin one construction and then quickly abandon it for another and then another?

■ Are students familiar with some of the shape names? Do they notice some of the characteristics of different shapes? Do they notice similarities and differences between shapes?

■ Can students count a set of 20 objects? How do they go about counting? Can they count accurately? Do they know the number sequence? Do they have a sense of the size of quantities up to 20?

Investigation 2: Exploring Numbers

■ How do students compare two numbers? Do they count out objects for each number and then compare the sets? Do they know which numbers represent larger quantities?

■ Can students match numbers and quantities? Can they use numbers to show how many? Can they order a set of numbers?

■ How do students combine two numbers? Do they count out two sets of objects and then combine them? Do they count on their fingers? Do they count orally starting from 1? Do they count on from one of the numbers? Do they use knowledge of number combinations?

■ How do students generate combinations of numbers equal to a given total? Do they use strategies based on counting? strategies based on number combinations? Do they seem to work strategically, or do they use trial and error?

■ How do students record their solutions to problems? Do they use pictures? numbers? words?

Investigation 3: Patterns

■ Can students describe a pattern that they see? Do they notice where it begins to repeat?

■ Can students predict the next element in a pattern? Can they do this for an a-b-a-b pattern? Can they predict what comes next in a pattern with three elements in each pattern unit, such as a-b-c-a-b-c or a-b-b-a-b-b? Do students begin to see how a pattern can be broken up into repeating units? For example, given the pattern yellow-green-green-yellow-green-green, do they see that the basic unit is yellow-green-green?

■ Can students construct their own patterns? Can they create simple patterns that alternate two colors? Can they create patterns that have more than two elements in each pattern unit? Do they begin a clear pattern but then lose track of what comes next? How do they describe and record their patterns?

Investigation 4: Counting and Combining

■ How do students keep track of the size of a set of objects they are accumulating? Each time they add a new group of objects to the collection, do they recount the whole set to find out how many they have? Do they count on? use mental computation? organize the objects in some way to make them easy to count?

■ Do they recognize when they have accumulated close to a given total number of objects? Do they know if they have more than the given total? less? Can they determine how many more or less they have?

■ How do students solve a problem that involves combining several quantities? How do students record and represent their strategies and ideas on paper?

Investigation 5: Data About Our Class

■ When students make representations of data obtained from a survey, does their work show what the survey is about? In what ways?

■ How do students' representations show different categories of data? Can you tell which pieces of data belong in each category?

■ Do students' representations show how many are in each category? Do they use pictures? numbers? words?

■ What strategies do students use to find the total number of people who answered the survey? What strategies do they use to determine which categories have the most and least?

■ Can students make sense of the results of their classroom surveys? Can they explain their ideas to others?

Thinking and Working in Mathematics

If you use this unit at the beginning of the school year, take this chance to observe students' work habits and communication skills. Think about these questions as you decide which routines, processes, and materials will require the most support through the year:

■ How comfortable are students in various types of work situations? Are they able to work alone? with a partner? Do they participate in whole-group discussions?

■ How do students respond to Choice Time? Are they self-directed and able to make choices independently? Do students move comfortably between activities, or do they stick with a familiar and safe choice? Do students complete each activity, or do they move around quickly from activity to activity?

■ How do students interact with peers? Do they share ideas? share materials? work cooperatively? Or do they prefer to work independently?

■ Can students express their ideas orally? Who participates in discussions?

■ Do students have ideas about how to record their work, either with words, numbers, or pictures?

■ What types of materials or activities do individual students seem most (or least) comfortable with?

In the *Investigations* curriculum, mathematical vocabulary is introduced naturally during the activities. We don't ask students to learn definitions of new terms; rather, they come to understand such words as *triangle, add, compare, data,* and *graph* by hearing them used frequently in discussion as they investigate new concepts. This approach is compatible with current theories of second-language acquisition, which emphasize the use of new vocabulary in meaningful contexts while students are actively involved with objects, pictures, and physical movement.

Listed below are some key words used in this unit that will not be new to most English speakers at this age level, but may be unfamiliar to students with limited English proficiency. You will want to spend additional time working on these words with your students who are learning English. If your students are working with a second-language teacher, you might enlist your colleague's aid in familiarizing students with these words, before and during this unit. In the classroom, look for opportunities for students to hear and use these words. Activities you can use to present the words are given in the appendix, Vocabulary Support for Second-Language Learners (p. 153).

the names of numbers from *one* to *twenty*
Students have many opportunities to count, combine, and compare numbers to 20; as a basis for this work, they will need to know the number names and their sequence.

numbers, words, pictures When students solve problems, they are asked to show their work using numbers, words, and pictures.

the names of colors Students work with pattern sequences based on color, so they need to know the names for the colors of the interlocking cubes you are using. They also identify cubes by color in the data investigation, if they use cubes to represent different data categories.

compare, larger (bigger), smaller, total In the games Compare and Double Compare, students determine which of two numbers and which of two totals is larger than the other.

peas, carrots, blueberries Students solve How Many of Each? problems that involve different numbers of peas, carrots, and sometimes blueberries.

predict Students predict "what comes next" in a pattern sequence, and also "what will happen next?" in a story.

Multicultural Extensions for All Students

■ Students who know the counting words to ten in other languages might share these with their classmates. The whole class might learn how to count to three in languages other than English.

■ Similarly, explore the words for color in other languages. When students are predicting cube patterns in Investigation 3, they might learn to recite the colors of a simple cube pattern in another language and predict what comes next; for example: *rojo, azul, rojo, azul, rojo, azul...*

■ Some of the children's literature suggested for use with this unit offers a glimpse into other cultures. For example, the book *Ten Little Rabbits* by Virginia Grossman and Sylvia Long, listed for use in Investigation 3, celebrates the customs of ten Native American tribes (the Sioux, Tewa, Ute, Menominee, Blackfoot, Hopi, Arapaho, Nez Perce, Kwakiutl, and Navaho). Help students find the patterns depicted in the weavings, decorations, and other native crafts.

Investigations

Exploring Materials

What Happens

Session 1: Getting Started with Materials The class establishes routines for using, caring for, and storing cubes, pattern blocks, and Geoblocks. Each student uses one of these materials to make a building or design.

Sessions 2, 3, and 4: Exploring Materials Students explore calculators and briefly share some of their discoveries with the class. Then they continue exploring cubes, pattern blocks, Geoblocks, and calculators, making sure that they spend some time using each.

Routines Refer to the section About Classroom Routines (pp. 145–152) for suggestions on integrating into the school day regular practice of mathematical skills in counting, exploring data, and understanding time and changes.

Mathematical Emphasis

- Exploring mathematical materials and tools, such as pattern blocks, interlocking cubes, Geoblocks, and calculators
- Comparing and finding relationships among geometric shapes
- Counting up to 20 objects

What to Plan Ahead of Time

Materials

- Interlocking cubes: at least 30 per student (Sessions 1–4)
- Pattern blocks: 1 bucket per 6–8 students (Sessions 1–4)
- Geoblocks: 1–2 sets per classroom (Sessions 1–4)
- Tubs or shoe boxes for storing materials
- Small containers, such as plastic cups or yogurt cartons, to use as scoops
- Calculators: 1 per pair (Session 2); 1 per 3–4 students (Sessions 3–4)

  Note: If necessary, try to borrow enough calculators for Session 2. If you do not have access to enough calculators, you can work with small groups at some time during Sessions 2, 3, and 4.

- Blank letter-size paper (available throughout, as needed)

Other Preparation

- Familiarize yourself with interlocking cubes, pattern blocks, and Geoblocks. If you have not used these materials before, spend some time exploring them. See the **Teacher Note,** Talking About Shapes (p. 11), for more information on pattern blocks and Geoblocks.
- A set of Geoblocks contains 330 blocks of 25 different types. Separate each set into two or three roughly equal subsets. If you make three subsets, each set will have enough blocks for 2–3 students to use at one time. With two subsets, there will be enough in each set for 4–5 students. The easiest way to divide the blocks is to find two (or three) identical blocks and put one in each set. Parent volunteers, aides, or older student volunteers might help with this.
- Sign and date, then duplicate the family letter (p. 148) to send home after Session 2.
- If you plan to provide folders in which students will save their work for the entire unit, prepare these for distribution.

Getting Started with Materials

Materials

- Interlocking cubes (at least 30 per student)
- Pattern blocks (1 bucket per 6–8 students)
- Geoblocks (1 or 2 sets, each divided into 2 or 3 smaller subsets)
- Containers for scooping cubes or blocks

What Happens

The class establishes routines for using, caring for, and storing cubes, pattern blocks, and Geoblocks. Each student uses one of these materials to make a building or design. Their work focuses on:

- exploring the characteristics of cubes, pattern blocks, and Geoblocks
- exploring geometric relationships by using geometric shapes
- using informal language to describe geometric shapes

Activity

Building Things

Begin by briefly introducing the three materials.

As we learn about mathematics this year, we will be using lots of different things to help us solve problems. We will be using tools like these cubes and blocks as we solve mathematics problems and play mathematical games.

Show students the interlocking cubes, the pattern blocks, and the Geoblocks. Ask students if they are familiar with any of these materials, how they have used them in the past, and if they know their names. Your students may not know many conventional mathematical terms for shapes; at this point, do not expect students use words like rectangle, hexagon, or pyramid. As they hear you and some of their classmates use them, they will begin to learn them naturally, as they learn other vocabulary. For more information on how first graders typically describe geometric shapes and how to talk about shapes with your students, see the **Teacher Note**, Talking About Shapes (p. 11).

With the introduction of any new material, it is important to establish clear ground rules. You will want to discuss where the materials are stored and how they will be used and cared for. The **Teacher Note**, Materials as Tools for Learning (p. 10), offers hints about establishing routines for using and caring for the manipulatives you'll be using.

For the next few days, you will be using pattern blocks, Geoblocks, and interlocking cubes. Each of these materials will be in a different part of the classroom. You'll need to plan your time so that you have a chance to use each of the materials. You may find out that one material is your favorite, but I'd like everybody to work with each of these—the cubes, the pattern blocks, and the Geoblocks—at least once during this time.

Set up the three materials in three different locations. You could use tables, clusters of desks, or rug space as places for students to work. Explain how many students can be at each "center" at one time. Depending on how you've organized your classroom, this could be indicated by the number of chairs at a certain table, or by posting the information on the board.

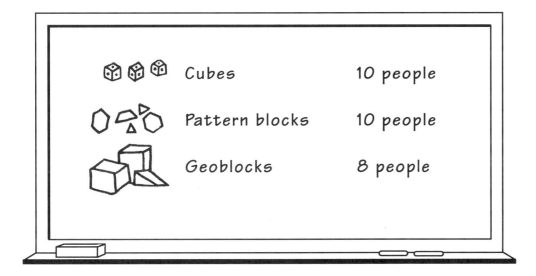

Ask students what material they would like to use today and arrange to have an appropriate number of students at each center. If you have 30 or more minutes left in the session, students may have time to explore two materials today. Otherwise, they will probably have time to explore only one material. Allow most of the class period for students to explore the materials. At the end of this session you will gather students together for a short discussion about their work (What Did You Notice? p. 7).

Suggestions for Using and Sharing Materials Many teachers decide to limit the amount of each material students can use. For example, they might ask students to take only a certain amount, or to build on a mat of a certain size. Constraints help students share limited materials. Having enough materials for all students to do what they want is always an issue. Some students may want to build things or create patterns that use an enormous number of cubes or pattern blocks, leaving other students without enough.

Constraints can also have mathematical value. Making something within a certain area or with a certain number of blocks leads students into many problem-solving situations. As students bump up against limits, they have to rethink their building or their pattern and make decisions about how to revise their work in a way that fits the constraints.

Constraints don't need to be too rigid. For example, if the way a student's pattern is developing makes it a little bigger than the mat, or if a student needs a few extra cubes to finish a building, it's fine to allow this.

You will need to decide what kinds of constraints make sense, given the availability of materials and the number of students in your class. Here are some possibilities:

■ **Interlocking cubes** Provide a small can or box to be used as a scoop. Each student can take one scoop of cubes to use (about 30–40 cubes is a good amount). Students may trade cubes, one for one, to get different colors if they need them, but can't increase their total number of cubes.

■ **Pattern blocks** Give each student a sheet of letter-size paper or cardboard to use as a mat. Any design must fit within the borders of the mat. If you have cardboard that can be used for mats, pattern block designs can be moved carefully to a display area at the end of the session, so that other students can see them. (Usually, pattern block designs can be kept only until the end of the school day, since the materials will need to be reused. Students need to understand this.)

■ **Geoblocks** Decide on a limit for the number of blocks students can use, or divide the Geoblocks into smaller sets to be used by two or three students at a time. Sharing will likely be an issue with this material since there are only a very few of some kinds of blocks. Discuss with the class whether "trades" between groups are allowed.

Throughout the year, make these and other mathematical materials, such as calculators, available for free play (for example, during an indoor recess). This will give students more time to explore the materials and will allow individual students or small groups to construct more elaborate buildings or patterns. See the Teacher Note, Supporting Students' Free Play (p. 8), for more information about how first graders use these materials and how to help students explore them.

Observing the Students

Use this time to gather information about how your students work with the materials and learn what ideas about geometric shapes they bring to their work. The section About the Assessment in This Unit (p. I-21) lists questions you can ask yourself about students' understanding.

Here are some things to look for as you watch your students at work:

■ Do they stick with the same material for a period of time or do they move from place to place?

- Do they work alone or in pairs? Do they talk to others about what they are doing?
- How do they describe the interlocking cubes? How do they describe the different pattern blocks: by shape or by color? How do they describe the different Geoblocks? Do they describe differences in size? Do they notice that some of the faces (they will probably call them "sides") are squares, some are rectangles, and some are triangles? What words do they use to talk about these shapes?
- Are any of their constructions symmetrical? Do they talk about symmetry in any way, for example, by saying that their pattern is "the same on both sides" or that parts of their constructions "match" or "balance"?
- With both the Geoblocks and pattern blocks, do they have a sense that pairs or combinations of blocks can be substituted for other blocks? For example, do they substitute two trapezoids for a hexagon, or put two Geoblocks together to match another Geoblock?

What Did You Notice?

About 15 minutes before the end of class, ask the students to clean up their materials, check the floor for stray cubes or blocks, and return all materials to their containers. Many teachers have found that giving a warning a few minutes before the end of a work period helps students be aware that a transition is about to occur. Establishing such routines is emphasized throughout this introductory unit.

Spend a few minutes asking students to share what they noticed about the materials. This discussion can be quite open. To get it started, you might ask questions like these:

Who can tell us something you noticed about the pattern blocks? Who noticed something different?

Who can tell us something you noticed about the cubes? Who noticed something else?

This discussion can be very brief, but it will give you some ideas about how students describe shapes. You might want to have a set of each material within your reach so you can hold up shapes that students talk about.

❖ **Tip for the Linguistically Diverse Classroom** For students who cannot communicate their observations in words, encourage them to show their thoughts, for example, by holding up two different Geoblocks and pointing to a face that is the same shape and size on both.

When students are first introduced to new manipulative materials, they make a wide variety of choices about how to use them. Many students are able to set themselves tasks and successfully create something that they are pleased with. However, while some students dive right in, others need some support and structure to help them take their first steps. When students are hesitant, help them notice what their classmates are doing. You might encourage any of these common approaches:

Making Buildings Sometimes students will start with an idea and carry it through: "I'm going to build a castle." Other times, they will just start building and then notice that their construction reminds them of something: "It looks like a Chinese temple. Now I'm going to make the steps."

Sometimes students working independently will join their separate buildings into one structure: "I made this part and Nadia made this part of it, and we decided to put it together. This part's the museum and this part is the living room, and the rest is everything else." Encouraging this kind of connecting of separate structures can also help when you are dealing with limited quantities of materials.

Accumulating Lots of Blocks Young students love to accumulate things. One building approach that reflects this love of accumulation is to enclose space in some way and then collect other blocks inside the enclosure: "This is the treasure room and these are the treasure." And: "This is a junkyard and this is where they store the tires, and this is where they have things that don't work any more, and these are the flags that are hung up."

Students like to make long trains or snakes with the cubes, or to build tall towers. Counting can come up with any of these materials, but it is especially likely with the interlocking cubes. Questions such as these lead to a focus on counting: "How tall can you make it before it breaks? Can you make a train as long as the table? How many cubes did you use?"

Making Patterns The pattern blocks, with their built-in geometric relationships, naturally lead to making pattern sequences or symmetrical designs. Some students use pattern blocks on their edges to make a "wall" in a certain pattern; for example, hexagon, triangle, hexagon, triangle, hexagon, triangle. Others make flat patterns, often working outward from a central hexagon.

Starting with a hexagon is something you might suggest to students who are stuck. Making a "wall" around a "garden" (which can be a piece of paper) is another possibility to suggest. Some students will also become intrigued with the symmetry of the Geoblocks and may make symmetrical constructions with them. This is a good opportunity to introduce the words *symmetry* and *symmetrical*: "Yours is symmetrical. It's the same on this side as on this side."

Sharing Good Ideas It's impossible to predict what will come up in your class, but be alert to ideas from your students that others in the class might enjoy. Sometimes these ideas sweep through the class without any help from you. For example, in one class, a student started using the interlocking cubes to make her initials. Soon other children were trying to make their own initials. The teacher engaged them in conversation about which letters were difficult and why. In another classroom, students (and the teacher) became intrigued with ways in which they could balance some of the Geoblocks on each other.

Establish Constraints It helps some students to get started if you set them a small problem with some constraint, either by limiting the number of blocks or cubes they can use, or by limiting the space on which they build. The problem can be quite simple; for example:

Make a design or building with a certain number of blocks.

This is similar to the Teacher Checkpoint activity, Counting 20 (p. 17), but can be done with any of the materials. You can adjust the number for different students. For example:

Make a design with exactly 12 pattern blocks.

Make a building with exactly 30 Geoblocks.

Here's another simple constraint:

Make a building or pattern that fits on your mat.

Using a letter-size sheet of paper as a mat helps students contain their work and limits the number of blocks or cubes anyone can use.

Describing and Recording the Work As students become more familiar with the materials, begin to put more emphasis on describing and recording both the individual blocks and the buildings and patterns students make with them. Ask students to notice and describe relationships between blocks. Inevitably, students will notice some combination of pattern blocks, such as two red trapezoids, that makes the hexagon shape. This is an opportunity to start them on a search for other combinations of blocks that make the hexagon. How many can students find?

Recording constructions can be difficult for some students, but very satisfying for others. Although students of this age have a range of developing fine motor skills, don't assume students can't record their buildings. Many can and will want to. For example, in one classroom, some students drew their Geoblock constructions in order to have a record of what they did. Other students have drawn pattern block patterns by tracing their designs.

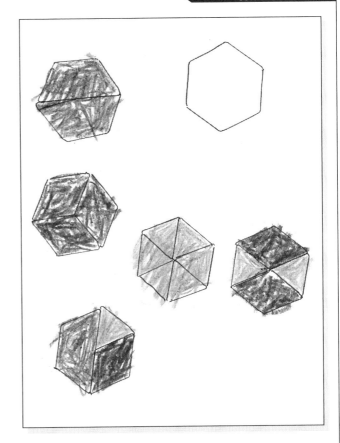

Some teachers run off pattern block shapes on colored paper and ask parents to cut up supplies for the classroom. You might keep a large envelope for each shape, with the shape pasted on the front. Parents take home sheets of the shapes whenever they can, cut them out, and drop them into the envelope when they are finished. Students can glue these shapes on white paper to record their designs. Commercial pattern block stickers can be used the same way.

Concrete materials are used throughout the *Investigations* curriculum as tools for learning. Students of all ages benefit from being able to use materials to model problems and explain their thinking.

The more available materials are, the more likely students are to use them. Having materials available means that they are readily accessible to students and that students are allowed to make decisions about which tools to use and when to use them. In much the same way that you choose the best tool to use for certain projects or tasks, students also should be encouraged to think about which material best meets their needs.

It is important to encourage all students to use materials. If manipulatives are used only when someone is having difficulty, students can get the mistaken idea that using materials is a less sophisticated and less valued way of solving a problem. Encourage students to talk about how they used certain materials. They should see how different people, including the teacher, use a variety of materials in solving the same problem.

Using concrete materials in the classroom may be a new experience for your students. Before introducing new materials, think about how you want students to use and care for them and how they will be stored.

Introducing a New Material Students need time to explore a new material before using it in structured activities. By freely exploring a material, students will discover many of its important characteristics and will have some understanding of when it might make sense to use it. Although some free exploration should be done during regular math time, many teachers make materials available to students during free times or before or after school. Each new material may present particular issues that you will want to discuss with your students. For example, some first graders make guns with the interlocking cubes. Some teachers have rules that there are no guns in the classroom. Some students like to build tall structures with the Geoblocks. In some classrooms, the teacher specifies only certain places where tall structures can be made—for example, on the floor in a particular corner—so that when they come crashing down, they are contained in that area.

Establishing Routines for Using Materials
Establish clear expectations about how materials will be used and cared for. Consider having the students themselves suggest rules for how materials should and should not be used; they are often more attentive to rules and policies that they have helped create.

Initially you may need to place buckets of materials close to students as they work. Gradually, students should be expected to decide what they need and get materials on their own.

There should be a cleanup routine at the end of each class. Most teachers find that stopping five minutes before the end of class gives students time to clean up materials and double-check the floor for any stray items.

Storing Materials Store manipulatives where they are easily accessible. Many teachers use plastic tubs or shoe boxes arranged on a bookshelf or along a windowsill. This storage can hold not only pattern blocks, Geoblocks, and interlocking cubes, but also calculators, counters (buttons, bread tabs, tiles), paper, and 100 charts.

Note: If you are using the full *Investigations* first grade curriculum, 100 charts (10-by-10 grids containing the numbers 1–100 in sequence) are introduced in the second unit, *Building Number Sense*. Some teachers decide to post, early in the year, a 100 chart or another listing of the numbers 1–100 (such as a number line), so that students come to recognize the chart and the numbers as a familiar part of the classroom environment. Other teachers decide that since so many new materials and methods are introduced in the first unit, they will not post a 100 chart until it is used in the curriculum.

Talking About Shapes

As you observe students working with cubes, pattern blocks, and Geoblocks, you can learn a lot about what they notice about shapes: what characteristics they attend to, what relationships they recognize, and what distinctions they make. For example, you might hear students say things like this:

"Hand me that little square block."
"I'm using diamonds all around the edge."
"Look, three of these blue ones can fit right on top of the yellow one."

At this age, students will not use many conventional mathematical terms. For example, they will probably not know that a blue pattern block is a *rhombus* or *parallelogram*, but may instead use the everyday term *diamond*. They are likely to know some geometric names, such as *square*, *circle*, and *rectangle*, but may sometimes apply these incorrectly. For example, they might call a cube a *square* or call a triangle a *rectangle*.

Enter into students' conversations, often using the same terms they are using, but also asking them questions or making comments that challenge them to be clearer and more precise. For example:

Claire: I'm using diamonds all around the edge.

Are you going to use all blue diamonds, or are you going to use some of the tan diamonds?

You can also use interactions to introduce conventional mathematical names for two- and three-dimensional shapes, so that students hear these terms used in context:

Luis: Look, three of these blue ones can fit right on top of the yellow one.

Luis noticed that three of these diamonds can fit right on top of the hexagon. Did anyone notice any other pieces that can fit together on the hexagon?

You don't need to insist that students use the conventional terms. They will begin to learn them naturally, the same way they learn other vocabulary—by hearing them used correctly in context. In geometry activities in later units, the students will have many experiences in classifying, describing, and defining shapes.

For your own reference, the following shapes are included in the pattern block and Geoblock sets. Before reading these descriptions, try your own sorting of the shapes in each set—especially the Geoblocks, which are probably less familiar. Once you have sorted the blocks into groups that you think "go together," the descriptions may make more sense to you.

Pattern Blocks There are six shapes in the pattern block set. We treat these blocks most of the time as if they are two-dimensional shapes, even though they do have thickness. All the shapes are polygons: closed shapes with straight sides.

Yellow hexagon. A hexagon is a six-sided polygon. The yellow pattern block is a *regular* hexagon because all the sides and angles are equal.

Red trapezoid. A trapezoid is a four-sided polygon that has one pair of parallel sides.

Green triangle. A triangle is a three-sided polygon. The green pattern block is an *equilateral* triangle because all the sides and angles are equal.

Orange square. A square is a four-sided polygon with four equal sides and angles. (A square is also a rectangle.)

Blue rhombus and tan rhombus. These two shapes are both parallelograms, or four-sided polygons with two pairs of parallel sides. They are a special kind of parallelogram, called a rhombus, that has four equal sides.

Although we name the pattern blocks as if they are two-dimensional shapes (hexagon, square, triangle), they are technically three-dimensional shapes, because they have a third dimension (depth). So the square is not actually a square; it is technically a rectangular solid or rectangular prism with two square faces. The convention for pattern blocks is to talk about them as if they are two-dimensional shapes, focusing on their faces. However, if any of your students point out their three-dimensional characteristics (for example, that the yellow hexagon also has some faces that are rectangular), certainly encourage them to talk about these ideas.

Geoblocks There are five general kinds of shapes in the Geoblock set, and 25 different blocks. All the shapes are *polyhedra*: three-dimensional solid shapes with flat faces.

Rectangular prisms. Prisms have two opposite faces that are the same size and shape (congruent). All other faces, connecting these two opposite faces, are rectangles. In *rectangular* prisms, the two opposite faces are rectangles, so all six faces are rectangles. Most boxes are rectangular prisms. You can also call these shapes *rectangular solids*. There are 11 different rectangular prisms in the set; of these, 5 are *square prisms* and 4 are *cubes*.

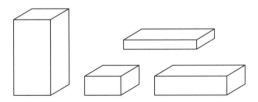

Square prisms. These are a special kind of rectangular prism. They have two opposite faces that are congruent squares. The other four faces are rectangles.

Cubes. Just as the square is a special kind of rectangle, the cube is a special kind of rectangular prism in which all the faces are squares.

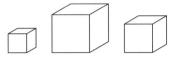

Triangular prisms. These prisms have two opposite faces that are congruent triangles. As in any prism, the faces that connect this pair are all rectangles. There are thirteen different triangular prisms.

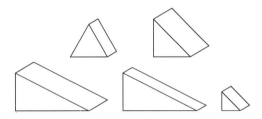

Pyramid. Pyramids look different from prisms. They have one base, which can be any polygon. The rest of the faces are triangles that meet in a single point (vertex). There is just one pyramid in the Geoblock collection, a *square pyramid*. It has a square base and four triangular faces.

Exploring Materials

What Happens

Students explore calculators and briefly share some of their discoveries with the class. Then they continue exploring cubes, pattern blocks, Geoblocks, and calculators, making sure that they spend some time using each. Each student makes a construction with exactly 20 cubes. Their work focuses on:

- exploring the calculator as a mathematical tool
- exploring cubes, pattern blocks, and Geoblocks
- using informal language to describe geometric shapes
- counting 20 objects

Materials

- Calculators (at least 1 per pair)
- Interlocking cubes (at least 30 per student)
- Pattern blocks (1 bucket per 6–8 students)
- Geoblocks (1 or 2 sets, each divided into 2–3 smaller subsets)
- Containers for scooping cubes or blocks
- Family letter (1 per family)

Activity

Exploring Calculators

You will need at least one calculator per pair for this activity; one for each student is ideal. If necessary, try to borrow some calculators from another classroom. If you do not have enough calculators, you might take turns doing this activity with small groups while the rest of the class is exploring the other materials during these three sessions.

We've been looking at some tools, like interlocking cubes and pattern blocks, that you can use to help you solve problems or to help you explain your thinking to others. Here's another kind of tool. *[Hold up a calculator.]* **Does anyone know what this is called? Who knows what we use it for?**

After a few volunteers share what they know about calculators, distribute at least one calculator to each pair. Give students about 10 minutes to explore them.

As you circulate around the room, you can get a sense of how familiar your students are with the calculator.

- Do they know how to turn it on?
- Do they recognize that the digits they enter appear on the screen display? Can they read numbers the screen display (in particular, any one- or two-digit numbers)? Do they notice numbers with decimal points?

- Do they know how to clear the screen?
- Are they familiar with the +, −, and = symbols? with any other symbols on the keyboard? Can they do any computations?

Your students may vary widely in what they notice about calculators and what they do with them. Some may use calculators only to make numbers; some may explore patterns, such as those created by repeatedly adding 2's; and some may make up and solve addition or subtraction problems.

Keep in mind that some students who can create and solve addition and subtraction problems with calculators (or even without calculators) may have little understanding of what it means to add and subtract, and little sense of whether the answer they obtain is reasonable. Do not insist at this point that students work only with operations they understand well or with small numbers. Often when students are first introduced to the calculator, they are fascinated by the power of being able to make large numbers or solve complicated arithmetic problems, even if they don't completely understand these. Students will have many opportunities throughout the year to develop their understanding of addition and subtraction; they will also learn to relate the processes of addition and subtraction to the symbols +, −, and =. The **Teacher Note**, Using the Calculator in First Grade (p. 20), contains suggestions on appropriate uses of calculators in your classroom.

Note: Students are likely to encounter decimals at some point in their calculator explorations. If students want to know what "the little dot" means, ask them first for their ideas. Some may suggest it's like a comma or period; others may think it signals that they have made a mistake. A few may know that it is used in money, and the occasional first grader will know that it's "the extra part" of the number.

Many first graders will not yet be ready to interpret the decimal portion as a part of a whole number. Explain to them that numbers with dots, or decimal points, are special kinds of numbers they'll learn more about in the next year or two. For now, when they encounter a number like 5.726, they can think of it as "about 5," or "5 and a little more," or "5 and some extra."

If you think a few students are ready to begin thinking about parts of whole numbers, you might explain to them that 5.726 is "between 5 and 6," as if you had 5 cookies and some pieces of another one, but not enough pieces to make another whole cookie.

Sharing What We Did Take about 10 minutes for a few volunteers to share something they noticed about the calculator or something they did with the calculator. The **Teacher Note**, Sharing Mathematical Ideas (p. 22), offers tips on having successful mathematical discussions with young students. As students share, listen for the chance to highlight the following:

■ the screen display, where digits appear

■ how to turn the calculator on and off

■ the numeral keys 0–9, and how to enter a two-digit number

■ how to clear the screen display

Show students how and where the calculators will be stored.

Activity

Buildings, Patterns, and Calculators

Explain to students that for the rest of today and most of the next two days, they will continue exploring calculators and the three materials introduced in Session 1 (interlocking cubes, pattern blocks, and Geoblocks).

Review the ground rules for using materials that you established in Session 1. You might also have a brief discussion about any issues that have been coming up. For example, if sharing has been a problem, you might want students to role play a situation that has come up in the classroom, and then talk about what works and what doesn't.

Before students get to work, check with them about which materials they have used and remind them they need to spend some time with each of the four materials. You may need to remind some students of this at the start of Sessions 3 and 4, and periodically during the sessions as well.

See the **Teacher Note,** Supporting Students' Free Play (p. 8), and the **Dialogue Box**, Mine Is a Sailboat (p. 23), for ideas about how to help students structure and focus their explorations, and for some particular challenges you can set for students who can't seem to get started or who appear bored. However, it's generally best if students explore the materials in their own ways for now, without a particular task set by the teacher.

During these three sessions, plan to work with small groups of students on the Teacher Checkpoint, Counting 20 (p. 17).

Since the students' work with manipulatives is not very portable, you may want to set aside time periodically when they can walk around the room and look at other work in progress. At the end of a session, before cleaning up, is one good time to do this. Some students may be interested in recording their work in some way, in order to preserve their ideas. For example, they might record their pattern block designs with stickers or cutouts, if you have them available. Even though this may seem a daunting task, some first graders are very interested in recording what they've made. See more about this in the **Teacher Note**, Supporting Students' Free Play (p. 8).

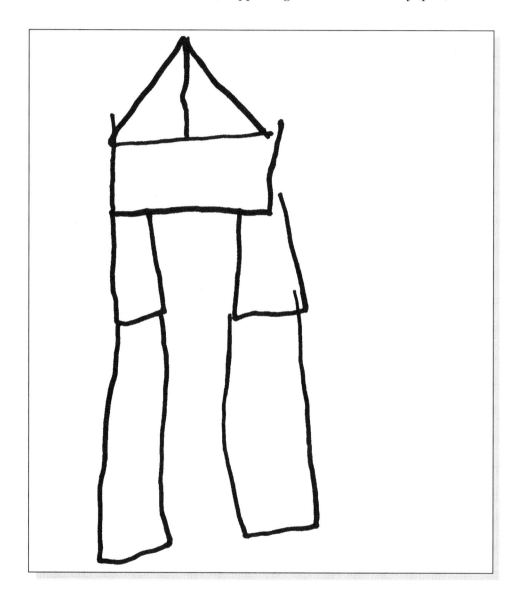

Teacher Checkpoints are places for you to stop and observe students at work. (For more information, read About Assessment, p. I-10.) Through this checkpoint, you can quickly get a sense of each student's counting and number sense.

Working with four students at a time, ask each one to make something with exactly 20 interlocking cubes. When a student is partly done, ask:

How many cubes have you used so far?

This will give you an opportunity to see the student count a quantity smaller than 20. To challenge students further, you might also ask:

How many more cubes do you need? Are you close to 20 yet?

When the students have finished, ask each one to prove to you that he or she used exactly 20 cubes.

How do you know there are 20? Can you show me that there are 20?

As students are working, watch for the following:

- Do students know the number names? Do they say the number names in sequence?

- Can students count accurately? Do they say one number word for each cube? Do they skip any cubes? Do they count any cubes twice? Can they keep track of which cubes they've counted?

- How do students keep track of how many cubes they have used so far? Do they recount the whole set each time they add a new cube? Do they remember how many cubes they have and say the next number in the sequence each time they add a cube?

- Do students have strategies for finding out how many more cubes they need? Can they tell you *exactly* how many cubes they still need? ("That's 17, so I need 3 more.") Or do they say *about* how many cubes they still need? ("That's 17, so I only need a few more.") Can they tell whether they're close to 20? ("That's 17, so I'm almost there.")

- Once they have counted out 20, are they confident that they have 20, or do they need to keep counting and recounting to be sure?

- Once their construction is finished, how can they show that they have used 20 cubes? For example, as they count, do they need to take the construction apart in order to keep track of which cubes they have counted? As they are counting, do they use any groupings to help them (for example, counting by twos)? Do they use any number combinations to help prove they have 20? ("I have 5 over here and 5 over here, so that's 10.")

You may find that some students have difficulty counting out exactly 20 objects; they may count some objects twice, skip one or more objects, or lose track of what they have counted so far. This checkpoint is intended only to give you a sense of students' skills as they enter first grade and learn the size of quantities with which they can work comfortably. For now, it is not necessary to correct students' mistakes or misconceptions. Your students will have many opportunities to develop their understanding of counting and numbers in this unit and throughout the year. The **Teacher Note**, Observing Students Counting (p. 19), explains further the development of counting abilities in the first grade.

Sessions 2, 3, and 4 Follow-Up

🏠 **Homework**

Family Connection Send home the family letter (p. 156) or the *Investigations* at Home booklet to encourage interest in your mathematics program and to let families know what to expect from the next few weeks of work.

Observing Students Counting

Your students will be counting many things during this unit and throughout the year. Counting involves more than knowing the number names, their sequence, and how to write them. It is the basis for understanding our number system and for almost all the number work primary grade students do.

In first grade, expect a great deal of diversity among your students. By the end of the year, many students will have learned the oral counting sequence up to 100 and will begin to recognize patterns in the sequence of numerals from 1 to 100. However, many first graders will not end the year with a grasp of *quantities* greater than 25 or so. Students develop their understanding of quantity through repeated experiences organizing and counting sets of objects. In first grade, many of the activities that focus on quantity can be adjusted so that students can work at a level of challenge that is appropriate for them. Early in first grade, some students will need repeated experiences with quantities up to 10, while others will be able to work with larger collections. Some students may be inconsistent—successful one time, and having difficulty the next.

Your students will have many opportunities to count and use numbers in this unit and throughout the year. You can learn a lot about what your students understand about counting by observing them as they work. Listen to students as they talk with each other. Observe them as they count objects and as they count orally and in writing. Ask them about their thinking as they work. You may observe some of the following:

■ *Counting orally*. Generally students can count orally further than they can count objects or correctly write numbers. For some students, the oral counting sequence is just a song; they don't necessarily know that when they count one more, they are referring to a quantity that has one more. Students need many experiences counting and adding small quantities as they learn about the relationship between the counting words and the quantities they represent.

■ *Counting quantities*. Some students may correctly count quantities above 20; others may not consistently count quantities smaller than 10. Some students may count the number of objects correctly when they are spread out in a line but may have difficulty organizing objects for counting themselves. They may need to develop techniques for keeping track of what they are counting.

■ *Counting by writing numbers*. Many beginning first grade students are just gaining some competency in writing numerals. Young students frequently reverse numbers or digits. Often this is not a mathematical problem but simply a matter of experience. Throughout the year, students need many opportunities to see and practice the sequence of written numbers.

Increasingly sophisticated calculators are used everywhere, from homes to high school classrooms to the workplace. If students are forbidden to use calculators in school while they see adults using them outside of school, they learn that "school" mathematics is nothing like mathematics in the real world. In the world around them, using a calculator is part of real life. We believe that students at all levels need to learn how to use calculators effectively and appropriately as a tool, just as throughout the elementary grades they learn to interpret maps, measure with rulers, and use coins.

Calculators enable students at all levels to apply their reasoning and problem-solving skills to a wider variety of problems. Students can combine the use of calculators with mental calculation or work with manipulatives as they solve problems that use large numbers or require many calculations.

For example, in the middle of the year in one first grade class, students were asked to determine how many crackers the teacher had brought in for snack: there were eight bags of crackers, and each contained six crackers. Students had been encouraged throughout the year to use a variety of tools to solve problems like this one. Of the many methods students came up with, several involved combining the use of the calculator with other mathematical tools:

■ One student recorded $6 + 6 + 6 + 6 + 6 + 6 + 6 + 6$, explaining that each "6" represents the number of crackers in one bag, and then used a calculator to find the sum.

■ Another student also began by recording $6 + 6 + 6 + 6 + 6 + 6 + 6 + 6$, but was able to solve more of the problem in her head. She knew that $6 + 6$ is 12, and recorded 12 under each pair of sixes:

$$6 + 6 + 6 + 6 + 6 + 6 + 6 + 6$$
$$12 \ + \ 12 \ + \ 12 \ + \ 12$$

She then used the calculator to add the four twelves.

■ Another student used the calculator to help her solve the problem by successive "doubling." She explained that since there are six in one bag, there are 12 in two bags. She knew that doubling 12 would give her the number in four bags, and doubling the number in four bags would give her the total in all eight. She then used the calculator to add 12 and 12 to get 24, and then to add 24 and 24 to get 48.

■ Yet another student recognized the problem as a multiplication situation. He knew that the problem could be represented by 6×8, but did not know how much that was. He used the calculator to find 6×8. Then, to check, he built a representation with eight towers of six interlocking cubes, and counted all the cubes.

When Should Students Use Calculators? For students at any level, there are times when it is appropriate to use calculators and times when it is important to develop and use other skills and strategies. You need to make decisions about when to allow students access to calculators.

You will probably find many situations in which using calculators can facilitate students' work with large numbers, as it did for the bags of crackers problem. Some of these situations may arise during math time, others outside of math. For example, in some first grade classes, students have used a combination of calculators and their own reasoning as one of several strategies for finding the total number of cans and bottles collected for a class recycling project, the amount of money collected at a class bake sale, and the number of cookies needed for a class party.

You will find many situations in which you do not want students to use calculators. When students are exploring number combinations with the How Many of Each? problems, and in later units when they are developing strategies for solving

addition and subtraction story problems, we suggest that students focus on strategies built on what they know about counting and number relationships.

As students progress through the elementary grades, they will need to begin making choices about when it is appropriate to combine calculators with their own reasoning and when it is appropriate to use other tools, such as mental calculation and estimation, paper and pencil, and manipulatives. It is important for first graders to begin learning about all these tools. To help them understand more about calculators and when it is appropriate to use them, give them opportunities throughout the year to freely explore and experiment wih them, just as they have many opportunities to "mess around" with other mathematical tools, such as pattern blocks, interlocking cubes, and paper and pencil.

Sharing observations and approaches to solving problems is one of the key ways in which students learn mathematics. Throughout the *Investigations* curriculum, students are encouraged to communicate about mathematical ideas in two major ways: by recording their work and by talking about it.

When students use pictures, numbers, or words to show how they solved a problem, they clarify their own ideas, learn how to justify their reasoning, and sometimes notice things that they hadn't observed when they were in the midst of doing the mathematics. By trying to represent or explain clearly what they are doing, they have a chance to reflect on what they did. Now that they are not engaged in doing the mathematics, they have a chance to view it as a whole, to think through again what they did, and to take away from the experience strategies and ideas that may be useful the next time they encounter a similar problem.

When students are called on to present their work to the whole class, they learn that their mathematical thinking is valued and they have the chance to develop confidence in their own ideas. Because this process of communicating about ideas is so central in learning mathematics, it is important to start establishing the expectation that students will describe their work even at this early age.

Keep in mind that students who are 6 or 7 years old cannot sit and listen attentively to this for very long. Since students are just learning to explain their work, their explanations are often hard to follow, even for you, and certainly for their classmates. You will need to balance the value of having students share their ideas with the difficulty that the class as a whole has with listening.

At first, keep whole-class discussions very short and focused, like the one suggested early in Investigation 1 (What Did You Notice? p. 7). Here students simply share what they noticed as they were engaged in an activity. The discussion is structured almost as list-making:

What did you notice about pattern blocks? Who can tell us something different?

Questions like these allow lots of students to participate without any one student taking a lot of time. You can also use questions that allow many students to participate at once by raising their hands. For example:

So Leah noticed that she could make a hexagon with two of the red trapezoids. Who can think of another way? Close your eyes and see a way in your head that you can make the hexagon with other blocks... OK, Brady, show us another way. How many people were seeing this way in their heads? Who was seeing a different way?

As the year goes on, students will become more used to listening to each other. Frequently ask students to listen to see if their approach is the same as or different from one that has been described. Since it is hard for young children not to have a turn when they have something to say, you might want to establish a "math show and tell" or a "math discovery of the day." This way every student gets a special turn to share from time to time, as well as the chance to participate in group discussions.

Mine Is a Sailboat

Students vary widely in the amount of structure and direction they need as they freely explore materials. Guiding students' exploration with questions can be an effective way of structuring a free exploration experience. This can extend students' thinking about a particular material, and lead them into new ways of using it. When students work with a partner or small group, they benefit from observing how others use materials. Inviting students to share their constructions and designs is a natural way for students to exchange ideas.

The following dialogue occurred early in the year during free exploration. The teacher first joins a small group of students working with pattern blocks and asks them to tell her about their constructions.

Donte: Mine is a sailboat. Here's the boat part, and then I made the mast with the skinny diamonds, and the sail is made with green triangles.

Iris: Mine is just a design. I started with this hexagon in the middle and then I built out and around. And now it looks like a pinwheel.

Your design is similar to Chanthou's design, but she started with two red trapezoids in the center.

Iris: It's the same thing, because she just used two of these [trapezoids] instead of this [hexagon].

William: Stop shaking the table! I don't want my wall to fall down.

I can see that some of you have decided to use the pattern blocks flat on the table and some of you have used them to build up, sort of like blocks. It is harder to keep them in place when they're on their edges, the way William has them. William, it looks like your wall has a pattern. Tell me about it.

The conversation is interrupted by some commotion at a table of students using interlocking cubes. Three students are building elaborate constructions and two are being silly, sliding cubes back and forth across the table. The teacher addresses these two students.

Tony and Michelle, have you figured out how to snap these cubes together so that they make a flat square? I noticed that yesterday when you were working with the Geoblocks, you were trying to build a very tall skyscraper. Do you suppose there's a way to use these cubes to build a tall skinny building?

Tony: Well you could put them together in a long line, like this. *[He snaps together a row of seven cubes.]*

Michelle: You could make it fatter by adding more on the bottom. *[She builds a 2-by-2 square and then begins to attach more cubes on top of her base.]*

I wonder how many floors your building could have? I live in an apartment build that has six floors.

Tony: My building has 11 floors. I'm going to make a building with 11 floors.

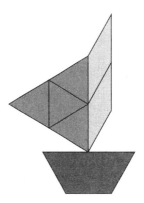

INVESTIGATION 2

Exploring Numbers

What Happens

Session 1: The Game of Compare Students play the game Compare, in which they find the larger of two numbers. As students become ready for more challenge, either in this session or later in the investigation or unit, they play the game Double Compare, in which they find which of two totals is greater.

Sessions 2 and 3: Introducing Staircases and Choice Time Students are introduced to a new activity, building "staircases" from interlocking cubes. Then, they participate in Choice Time for the remainder of the sessions. The choices include building staircases and playing Compare or Double Compare.

Session 4: Seven Peas and Carrots Students solve a problem in which they have seven peas and carrots altogether. They determine one or more combinations of peas and carrots they could have to make up seven in all. This is the first of several How Many of Each? problems they will solve throughout the school year.

Sessions 5 and 6: Number Choices During Choice Time, students work on the following choices: solving a How Many of Each? problem involving nine peas and carrots, building staircases, and playing Compare or Double Compare. At the end of Session 6, they share ways that they recorded solutions to the How Many of Each? problem.

Routines Refer to the section About Classroom Routines (pp. 145–152) for suggestions on integrating into the school day regular practice of mathematical skills in counting, exploring data, and understanding time and changes.

Mathematical Emphasis

- Developing strategies for comparing two quantities up to about 20
- Using numbers to show how many
- Developing strategies for combining two single-digit numbers
- Finding combinations of numbers up to about 10
- Representing solutions to problems with pictures, numbers, and words
- Ordering a set of numbers up to about 20
- Counting up to 20 objects

What to Plan Ahead of Time

Materials

- Chart paper or newsprint (18 by 24 inches): 15–20 sheets (available for teacher use)
- Blank letter-size paper (available for student use)
- Interlocking cubes: at least 30 per student
- Number Cards: 1 deck per pair, stored in resealable plastic bags or envelopes. If you do not have manufactured cards, make your own; see Other Preparation. (Sessions 1–3)
- Counters, such as buttons, bread tabs, or pennies: at least 30 per student (available for student use in all sessions). Be sure you have some counters in different colors.

Other Preparation

Duplicate the following student sheets and teaching resources, located at the end of this unit. If you have Student Activity Booklets, copy only the items marked with an asterisk.

For Session 1

Student Sheet 1, Compare (p. 157): 1 per student (homework)

Number Cards (pp. 174–177): 1 set per student (homework). If you do not have manufactured cards, you will also need a class set for each pair. Class sets will last longer if duplicated on card stock. Cut apart each set and store in a plastic resealable bag. A parent or other volunteer might help with the cutting.

For Sessions 2 and 3

Student Sheet 2, Double Compare (p. 158): 1 per student (homework)

Staircase Cards* (p. 160): 3–4 sheets, preferably on card stock. Cut apart each sheet to make two sets of 1–12 cards and two sets of 13–20 cards. (Use of the 13–20 set is optional.) Store each set in a plastic resealable bag.

Game Record Sheet* (p. 178): 1 per student (homework, optional)

For Sessions 5 and 6

Student Sheet 3, Bats and Balls (p. 159): 1 per student (homework). Consider modifying the number on this sheet for some students.

The Game of Compare

Materials

- Number Cards, with wild cards removed (1 deck per pair, and 1 set per student for homework)
- Student Sheet 1 (1 per student, homework)
- Interlocking cubes (class set)
- Counters (such as buttons or bread tabs)

What Happens

Students play the game Compare, in which they find the larger of two numbers. As students become ready for more challenge, either in this session or later in the investigation or unit, they play the game Double Compare, in which they find which of two totals is greater. Their work focuses on:

- playing a mathematical game with a partner
- comparing two numbers to find which is larger
- combining two quantities

Activity

Compare

Compare is a number card game that students play in pairs. Introduce this game to the entire class by assembling students in a circle on the floor to watch a demonstration game, with either two student volunteers or you and a student as the two players. Have a deck of Number Cards with the wild cards removed.

Today we will play a game called Compare. At the beginning of the game, each player gets half the cards in the deck.

Demonstrate how to deal out the cards evenly between the two players. One student takes the deck and gives a card to his or her partner, and then takes a card. The student continues giving one card away and taking a card until all the cards have been distributed. (Each player should have 22 cards.)

You will turn your cards facedown. Then, you will both turn over your top card. If your number is larger than the other player's, you say "Me!" Let's see what happens when our volunteers do it.

After the volunteers turn over their first cards, ask the class which number is larger and how they know. Pause for a moment to ask students which card is larger and how they know.

Tuan turned over 7, and Tamika turned over 4. Which is larger? Who says "Me"?

Play two or three more turns, or until you think students understand the game.

Sometimes you might each have the same card. When that happens, nobody says "Me." You just both turn over the next card. The game is over when both of you have turned over all your cards.

Note: If your students are familiar with the card game War, they will probably need just a brief introduction to Compare. An important difference is that in Compare, players do not win or lose because they do not capture each other's cards. Students who are familiar with War may naturally want to capture the other cards when their number is larger. You can decide how important it is for students to play less competitively; if the competition is distracting students from thinking about the numbers, you may want to insist that they play according to the Compare rules.

As students pair up to play the game, give each pair a deck of Number Cards with the wild cards removed.

Observing the Students

Circulate to observe how students are playing the game and to offer support as needed. Here are some areas in which to focus your observations:

■ Can students deal out the cards evenly between two players? (You might suggest that the students repeat "one for you and one for me" while dealing to help them remember how to distribute the cards.)

■ Do students understand the rules of the game?

■ Can they read and interpret the numerals on the cards, or do they count the pictures on the cards to figure out the number? Can they count the pictures accurately? If some students are having difficulty distinguishing 6 from 9, remind them to count the pictures on the cards to check, and to orient the cards so that the number is on the top.

■ What strategies do students have for determining which number is larger? Do they just "know" which number is larger? Do they count pictures on the card to help them? Do they use counters, such as interlocking cubes, to help them?

■ Do students play cooperatively? If you think they are playing too competitively, emphasize that getting the larger number is a matter of luck, not a matter of being a "better" player. Explain that good players play cooperatively, for example, by checking or helping one another, by explaining their thinking to one another, by asking one another for help, or by waiting while the other player takes the time to determine which number is larger.

If some students are having difficulty comparing numbers, you might suggest that each student in the pair take as many interlocking cubes as the number on his or her card and make a stack. They can then compare the heights of the stacks. The student with the higher stack says "Me."

You might also call together a small group and work with them as they play Compare. As they play, encourage them to work slowly in order to find and share ways to count and compare.

After students have played a few games, you might decide to introduce the following variation to some pairs: The player with the smaller number says "Me." (Note that the other two variations listed with the game directions on Student Sheet 1 are more difficult and should not be introduced until students are more familiar with the game.)

Double Compare

Double Compare is a variation of Compare; each player turns over two cards and finds the total of those numbers. Then the two players compare totals to determine which is greater.

When to Introduce the Game Introduce Double Compare as students become ready for a game more challenging than Compare. Some may be ready after playing Compare a couple of times; others may benefit from more experience with Compare. The **Teacher Note**, First Graders: A Wide Range of Understanding (p. 33), discusses the need to adjust most activities throughout the year for the wide range of mathematical understanding in your class.

If only a few students seem ready for more challenge at this time, introduce Double Compare to a small group or to pairs as the rest of the class continues playing Compare. You can continue teaching the game to small groups or to pairs as they become ready later in the investigation or unit. If most or all of your students are ready for more challenge, introduce Double Compare to a larger group. Whenever you decide to introduce the game, make sure that there is time left in that session for students to play a round or two after your introduction.

Introducing the Game Play a demonstration game with a student to introduce the rules.

[Demonstrate.] **I turned over 2 and 9. What's my total?**

For a whole-class demonstration, you might quickly sketch on chart paper each pair of cards that is turned over. Include both the numbers and the pictures on the cards (you could draw circles or dots in place of the pictures). Give students a few minutes to find the total of 2 and 9. They may count the pictures on the cards or use counters as needed.

Susanna turned over 5 and 5. What's her total? How do you know? *[Again, give students time to find the total.]* **My total is 11, and Susanna's is 10. Which is greater? How do you know?**

As in Compare, when players have the same total, they simply go on to the next two cards. The game is over when players have turned over all their cards.

Observing the Students

As students play Double Compare, watch for the following (see also the questions suggested for Compare, p. 26):

- How do students combine the two quantities? Do they count all the pictures? count on from one number? "just know" the number combinations?

- What strategies do students use for determining which total is greater? Students may sometimes be able to decide which pair is more without actually figuring out the total.

The **Teacher Note**, Double Compare: Strategies for Combining and Comparing (p. 30), describes some of the many strategies students might use to play the game and suggests ways to adjust the game to make it more appropriate for students at different levels.

After students have played a few games, you might introduce the variation in which the player with the smaller total says "Me." (As for Compare, the other two variations listed with the game directions should not be introduced until later.)

Near the End of the Session When 5 or 10 minutes remain in the session, review cleanup procedures, including how to return materials to storage areas and how to double-check the floor for pencils, stray counters, and other materials.

Session 1 Follow-Up

Compare Send home Student Sheet 1, Compare, and a set of Number Cards for students to use as they teach someone at home to play Compare. (Students can cut part the cards at home; if they are unlikely to have scissors, give them time during the school day to do this.) You may want to provide an envelope for storage of the cards. Advise students to keep their cards in a special place at home, because they will be playing card games for homework throughout the unit. Some teachers give students a "Math at Home" folder to use for storing the materials they will be bringing home throughout the year.

 **Homework**

Double Compare: Strategies for Combining and Comparing

Through the game of Double Compare, students develop their strategies for combining two numbers and for reasoning about quantity. The following scenes from a classroom illustrate situations that commonly arise, and show how to adjust the game for students at different levels.

Counting Objects

As Garrett turns up 3 and 0 and Shavonne turns up two 8 cards, Garrett begins by reminding himself, "Count those little things [the pictures on the cards]." Then, while Shavonne watches, he counts each picture on the 3 card, touching them as he says the numbers. He announces that he has 3.

Shavonne places her cards side by side, overlapping the edges. She counts slowly, touching all the pictures as she says the numbers, but she skips a few pictures, counts a few twice, and comes up with a total of 13. Garrett says that he thinks 8 and 8 is 18. Although they are aware that at least one of these totals is inaccurate, they realize that regardless, Shavonne's total is greater than Garrett's total of 3, and they are ready to move on.

At this point the teacher steps in and asks them to recount Shavonne's total, slowly. After a couple of trials, Shavonne and Garrett both come up with a total of 16. The teacher suggests that they use interlocking cubes to help them find the totals on their cards. Since cubes, unlike the pictures on the cards, can be moved around, they can make it easier for students to keep track of what they have counted and what they have left to count.

Shavonne and Garrett both need to count by ones to be sure of their totals, and counting totals greater than 10 is challenging for them. The teacher plans to return in a few minutes to see if the cubes are helpful. If Shavonne and Garrett are still having difficulty working with larger numbers, she will suggest that they play with only the 1–6 cards. Later in the session, she will call together students having difficulty and will work with them as they play Double Compare.

Counting and Counting On

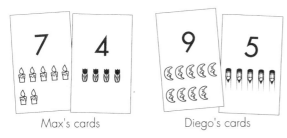

Max's cards   Diego's cards

Max and Diego get right to work finding their totals. Diego counts quietly to himself. He begins at 9, and then counts "10, 11, 12, 13, 14." With each number he says, Diego uses his right index finger to bend back one of the fingers on his left hand. When he has bent back all the fingers on his left hand, he stops counting and announces that he has 14.

Meanwhile, Max is still counting. He began by looking first at the 7 card and counting from 1 to 7. Then, he turned to the 4 card and began counting "8, 9...." When Diego announced his total of 14, Max lost his place. He begins counting again. He counts to 7, and then he counts the pictures on the 4 card, saying "8" as he points to the first picture, "9" as he points to the second, and so on, until he reaches 11. The boys agree that Diego has the greater total.

Like Shavonne, Max puts the two quantities together and counts them all, starting at 1. Diego can begin with one quantity and count on by starting at the next number. In order to do this, Diego treats 9 as a unit; that is, he can think of it as 9 without breaking it down into ones again. Then he counts on—"10, 11, 12, 13, 14"—while keeping track with his fingers of how many he needs to add (1, 2, 3, 4, 5). The teacher feels that the game is at an appropriate level of challenge for Diego and Max. In future sessions, she will observe them to see how their strategies for counting and combining are developing. For example, she will note whether Max continues counting from one each time, and whether they have begun developing strategies for determining particular combinations without counting.

"Just Knowing" Number Combinations

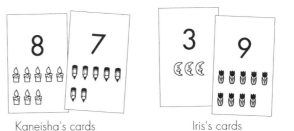

Kaneisha's cards   Iris's cards

As Kaneisha and Iris turn over their cards, Kaneisha immediately announces that she has 15, then looks over at Iris's cards. The teacher reminds Kaneisha to let Iris find her own total. Meanwhile, Iris counts almost inaudibly to herself "10, 11, 12," and then says that her total is 12. Kaneisha says, "Me! I won."

How did you get your totals?

Kaneisha: Because I know 8 and 7 makes 15. Because it's easy.

Iris: I counted in my mind.

Kaneisha's cards   Iris's cards

On the next round, Kaneisha again immediately announces her total, and then waits impatiently while Iris slowly counts on from 9.

When the teacher again asks how the girls found their solutions, Kaneisha is still unable to explain. She seems either to have memorized some number combinations or to have developed strategies for finding solutions to number combinations quickly. As Kaneisha is eager to play at a faster pace than she can with Iris, the teacher decides to ask her to play with Nathan, who is also finding number combinations quickly. To provide further challenge, she may ask Kaneisha and Nathan to turn over three cards on a round.

Reasoning About Number Combinations

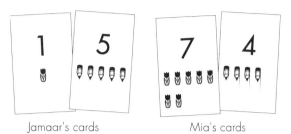

Jamaar's cards   Mia's cards

Jamaar: One and 5 is 6. I have 6.

Mia: 8, 9... *[after a short pause]* Me! Because you have 6 and I have more. Let's do it again.

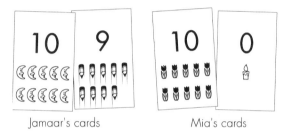

Jamaar's cards   Mia's cards

Jamaar: Me! 9 is bigger than 0. You know because it's just your eyes that tell you.

Mia and Jamaar are reasoning about the number pairs without necessarily needing to add them up. Although the teacher has observed in previous sessions that Jamaar and Mia are skilled at counting, comparing, and combining numbers, she feels that this game is deepening their understanding of numbers and number relationships as they explore ways to reason about numbers.

Choosing a Card to Win a Round

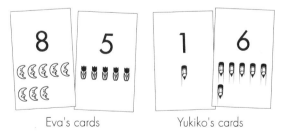

Eva's cards   Yukiko's cards

Eva and Yukiko count together slowly, starting with Eva's cards. Eva touches the pictures as she counts.

Yukiko tells her she has counted a picture twice, and they begin again. After several trials, they complete the count with a total of 12.

Yukiko: Six and 1 is 7. But it's my turn to win this time... *[She places her cards side by side so that the 1 is to the left of the 6.]* It's 16. I win!

They both laugh, knowing that the pictures can be combined to find the total number, but the digits cannot be combined in that way.

Next, Yukiko removes her 1 card and places it facedown in her discard pile. She pulls out a 3 card from her pile and then returns it, saying "Three, too small." Then, she pulls out a 2 card from her pile, hesitates, and puts it back, saying "I need something big." Finally, she pulls out a 9, and puts it face up with the 6.

Yukiko: There. It's bigger than your 12. I won.

Eva: OK. My turn to win next.

Eva and Yukiko have invented a version of Double Compare in which players win alternate rounds. If the player whose turn it is to win has a losing hand, she can choose a replacement card. Although Eva and Yukiko appear to need more practice counting and combining (they struggled to combine 8 and 5, and arrived at an incorrect total), the teacher decides not to intervene at this point. Although they are not always finding the total of their combinations correctly, they are nonetheless gaining practice counting, comparing, and combining. They are concentrating fully on their work and, as they find winning combinations, they are reasoning about the relative size of numbers and number combinations. The teacher compliments them on developing a collaborative version of the game.

First Graders: A Wide Range of Understanding

Students enter first grade with a very wide range of mathematical understanding and experience. At this age, they are in the midst of an important cognitive transition, from reliance on concrete modeling of mathematical situations to the beginnings of reasoning about numbers, quantities, shapes, and other mathematical objects. This transition lasts for several years as students slowly explore how to make mental models of important mathematical relationships. The 4- or 5-year-old who feels confident that "the numbers go on forever," even though she can't actually count past 29 without help, has developed a model of the number system that no longer depends on what she can see and touch in the world. She is beginning to build an abstract idea about what numbers are and what they do. The 6-year-old who says, "I know that 8 and 5 is bigger than 7 and 5 because 8 is bigger than 7" is reasoning about number relationships. He knows something about the operation of addition and how it works; he realizes that even without figuring out the two sums, he can tell how the two sums relate to each other.

Understandings like these come slowly. Students need many experiences with counting, comparing, building numbers out of cubes or tiles, solving problems, drawing pictures, and talking about their strategies as they develop ideas about the number system and about operations with whole numbers. Some students entering first grade will not be able to reliably count small quantities. Others can easily count or recognize small quantities without counting, but will have more trouble counting quantities greater than 12 or 13. Some students are confident counters, but will rely completely on counting by ones to solve any problem. Other students are beginning to reason about numbers and operations: "I know that 6 plus 1 is 7, because 7 is the number that comes after 6." And: "I know 6 take away 3 is 3 because 3 plus 3 is 6."

Because students' ideas about number can develop at very different paces throughout first grade and well into second grade, you will need to pay a great deal of attention to individual needs and modify many of the first grade activities, so each student is working on the mathematical ideas most critical for him or her. You will want to make sure students have enough challenge but are not overwhelmed. The **Teacher Note**, Double Compare: Strategies for Combining and Comparing (p. 33), illustrates how a teacher can make use of the suggestions in this unit to modify activities as needed. Like Double Compare, many of the games and activities in the grade 1 *Investigations* curriculum are designed so that students working on different aspects of a mathematical idea can participate fully.

Modifying activities for different students is often as simple as changing the numbers being used, or changing the number of turns, or some other variable (like the number of cards used, turning Compare into Double Compare or Triple Compare). Specific suggestions, based on the variations that have been tried in classrooms, are included in each unit, along with examples of student work you can use as a guide to match variations to your students. You will also need to make longer-term decisions. For example, some students may need to continue playing Compare throughout this unit, while others are ready to play Double Compare right away. Only your own careful listening, observing, and questioning will provide the information you need to help each student work on significant mathematical ideas at the right level of challenge.

Introducing Staircases and Choice Time

Materials

- Class sets of Number Cards (with wild cards removed)
- Counters (buttons, bread tabs)
- Interlocking cubes: at least 30 per student
- Staircase Cards (1–12): 6–8 decks
- Staircase Cards (13–20): 6–8 decks (optional)
- Student Sheet 2 (1 per student, for homework)

What Happens

Students are introduced to a new activity, building "staircases" from interlocking cubes. Then, they participate in Choice Time for the remainder of the sessions. The choices include building staircases and playing Compare or Double Compare. Their work focuses on:

- combining two quantities
- comparing two numbers to find which is larger
- counting a set of objects
- using numbers to show how many
- ordering a set of numbers

Activity

Introducing Staircases

Gather students in a circle. Bring interlocking cubes and a deck of Staircase Cards 1–12. Scatter the cards face up in the center of the circle so that students can see them. Make sure the cards are not in order. Ask for a few volunteers to tell what they notice about the cards.

What numbers do you see? What's the smallest number? the largest number? Do the cards show all the numbers from 1 to 12?

Count from 1 to 12 with the class, pointing to each number as you say it.

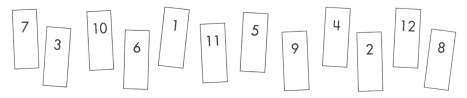

We're going to start making a staircase of cubes to match these cards. We'll start with 1. *[Put out one cube.]* **Here's the first step. We'll put the 1 card in front of the step to show that it has one cube in it.** *[Place the 1 card in front of the cube.]*

Now we'll build the second step. Will our next step have more cubes in it? How many more? Will it be higher than this one?

After a few students have offered their ideas, snap together and place a step of two cubes next to the first step; label it with the 2 card. Continue building steps until you think students understand the task.

Explain that today and tomorrow, everyone will have a chance to work with a partner to build a staircase of cubes. The smallest step will have one cube, and the largest will have 12. Students label each step with the matching card. Explain that students do not have to build and label one step at a time, as in your demonstration. They might instead build the entire staircase first and then label it with cards, or arrange the cards first and then build the staircase.

Activity

Introducing Choice Time

Choice Time is a format that recurs throughout the *Investigations* curriculum. See the **Teacher Note**, About Choice Time (p. 40), for information about setting up Choice Time and how first grade students can keep track of the choices they have completed. Tell students that during each day of Choice Time, they get to choose one or two activities to do. They can select the same activity more than once, but they also have to try more than one activity—they can't do the same one over and over, every day.

List the activity choices on the board or on a piece of chart paper, with a simple sketch as a nonverbal reminder.

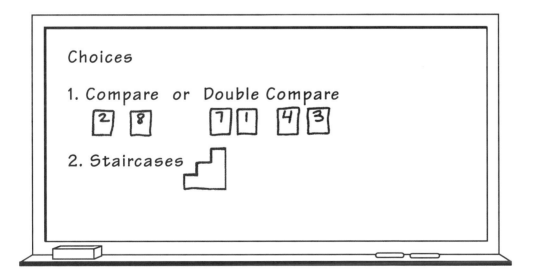

Note: Include Double Compare as an option for the first choice if you have already introduced the harder game to some students (or if you plan to do so during Choice Time). When you post the list, explain that some students have learned (or will be learning) a new version of Compare.

You will need to decide whether students who have been playing both games may choose between them. For example, some students who find Double Compare a little challenging may still want to play Compare occasionally. Many first graders will need your help in determining which game offers them an appropriate level of challenge.

You will also need to decide how many students at one time can work on Staircases, depending on your supply of cubes. Each pair will need at least 100–150 cubes. You will also need some cubes available for students who need help counting and comparing when playing Compare or Double Compare. For example, if you have 1000 cubes, you might decide that up to five pairs at a time can work on Staircases. That way, there will be plenty of cubes available for students to build staircases up to 12, and enough cubes for students who are ready for more challenge to build higher. You will also have plenty of cubes for students who need them for Compare or Double Compare.

Tell students how many pairs can work on Staircases and where they should get their materials (you may want to have a certain number of cubes available at a Staircase center). Assure them that some time over the next few sessions, everyone will have a chance to build staircases.

Choice 1: Compare or Double Compare

Materials: Decks of Number Cards, with wild cards removed (1 per pair); interlocking cubes or other counters (available)

Students will be familiar with Compare from their work in Session 1. Review the rules as needed (p. 26). When you think students playing Compare are ready for more challenge, introduce them to Double Compare (see p. 28).

If you notice students having difficulty, you might call together a small group and work with them as they play one of the games. Encourage them to go slowly in order to find and share ways to count, combine, and compare. Suggest using manipulatives or counting on their fingers. Students having difficulty with Double Compare might benefit from more practice with Compare, or from playing Double Compare with only the 1 to 6 cards.

Choice 2: Staircases

Materials: Interlocking cubes; Staircase Cards 1–12 (and 13–20, optional)

Students work with a partner to build a staircase of cubes, the smallest step having one cube and the highest having 12. They label each step with the matching Staircase Card.

For some students, the basic activity provides an appropriate level of challenge. They will need time to count and recount the number of cubes in each step, to order the steps and numbers, and to match various quantities of cubes with numerals. Other students may complete the task quickly and accurately, demonstrating a clear understanding of the size of quantities up to 12 and the numerals that represent those quantities. You might suggest one or more of the following to students who are ready for more challenge:

■ Building staircases higher than 12. Students can label the higher steps with Staircase Cards 13–20, or they can make their own Staircase Cards.

■ Building staircase "patterns." For example, students might build staircases that reflect the following:

Every other number, starting at 1 (the first step is 1, the second is 3, the third is 5, and so on)

Every other number, starting at 2 (the first step is 2, the second is 4, the third is 6, and so on)

Every third number, starting at 1 (the first step is 1, the second is 4, the third is 7, and so on)

Students may build up through the numbers in the teens, or higher if there are enough cubes.

■ Building "random" staircases. Students shuffle a set of Staircase Cards, 1–20, and select about six cards at random. They build a step for each card chosen and arrange them in order from largest to smallest.

Observing the Students

Observe and listen to students while they work on Choice Time activities. Recording your observations will help you keep track of how they are interacting with materials and solving problems. The **Teacher Note**, Keeping Track of Students' Work (p. 42), offers some helpful strategies.

During this first Choice Time, watch for the following:

■ Do students try each choice, or do they stay with a familiar one? If, after a short time with one activity, students say they're done, ask them to tell you about what they have done. Encourage them to investigate further.

■ How much do students interact with their partners? Do they share what they have done with others and observe what others are doing? Do they talk to themselves or others about what they are doing?

Compare and Double Compare

■ Can students interpret the numerals on the cards, or do they count the pictures on the cards? Do they count accurately?

■ What strategies do students have for determining which number is larger? Do they "just know"? Do they count pictures on the card? Do they use counters, such as cubes?

■ In Double Compare, how do students combine the two quantities? Do they count pictures? count on from one number? use their knowledge of number combinations?

■ In Double Compare, what strategies do students use for determining which total is greater? Do they just know? Do they use counters? Do they reason about number combinations?

Staircases

■ How comfortable and accurate are students as they count cubes? What sorts of errors do you notice in their counting? Are they able to keep track of what has been counted and what needs to be counted?

■ Do students recognize that each successive step will be larger than the next? Do they build each step by counting out each cube in a step, or do they build a step as high as the previous one and then add one more cube?

■ Can students order the cubes accurately? Can they order the numbers accurately?

■ Can students match the corresponding numbers and amounts of cubes?

Near the End of the Session Five or 10 minutes before the end of each Choice Time session, announce that it's time to stop working, put away the materials, and clean up the work area. A 5-minute warning before cleanup is a helpful way to signal that the activity is coming to an end and that students should be finishing up their work. Ask everyone to double-check the floor for pencils, stray cubes, and other materials. Once cleanup is complete, remind students to record the choices they have completed. See the **Teacher Note**, About Choice Time (p. 40), for suggestions on how your students might keep track. For example, they could write the names of their choices on their own lists , or write their own name on the list you have posted for each choice. If you have posted lists, keep them for further use in the Sessions 5–6 Choice Time.

Sessions 2 and 3 Follow-Up

Compare or Double Compare Students continue playing Compare at home, using the set of Number Cards they took home earlier.

 **Homework**

If some students have learned how to play Double Compare, they can now teach that game to a family member. These students should take home a copy of the directions (Student Sheet 2, Double Compare).

The Game Record Sheet (p. 178) can be used to help students reflect on the games they are playing at home. It also helps them understand that practice at home is an important part of the mathematics work they are doing. Sometimes when you are suggesting a game for homework, you may want to include a copy of this sheet for them to fill out (with family help) and return. Because this sheet can be used for any game, you will need to write the specific game at the top of the sheet before duplicating.

When you first send home this sheet, talk briefly about "telling the story" of what happened in a game. For example, one response after playing Compare might be "I used buttons to figure out the totals. Me and my father had two ties."

Triple Compare This game is played like Double Compare, but each player turns up *three* cards on a turn and finds the total. The player with the greater total says "Me."

 **Extension**

Choice Time is an opportunity for students to work on a variety of activities that focus on similar mathematical content. Choice Times are found in every unit of the grade 1 *Investigations* curriculum. These generally alternate with whole-class activities in which students work individually or in pairs on one or two problems. Each format offers different learning experiences; both are important for students.

In Choice Time the activities are not sequential; as students move among them, they continually revisit some of the important concepts and ideas they are learning. Many Choice Time activities are designed with the intent that students will work on them more than once. As they play a game a second or third time, or as they work to solve similar problems, students are able to refine their strategies, see a variety of approaches, and bring new knowledge to familiar experiences.

You may want to limit the number of students working on a particular Choice Time activity at any one time. In many cases, the quantity of materials available limits the number. Even if this is not the case, limiting the number is advisable because it gives students the opportunity to work in smaller groups. It also gives them a chance to do some choices more than one time. Often when a new choice is introduced, many students want to do it first. Assure them that, even with your limits, they will have the chance to try each choice.

Initially you may need to help students plan what they do. Rather than organizing them into groups and circulating the groups every 15 minutes, support students in making their own decisions. Making choices, planning their time, and taking responsibility for their own learning are important aspects of a student's school experience. If some students return to the same activity over and over again without trying other choices, suggest that they make a different first choice and then do the favorite activity as a second choice.

How to Set Up Choices

Some teachers prefer to have the choices set up at centers or stations around the room. At each center students will find the materials needed to complete the activity. Other teachers prefer to have materials stored in a central location, with students taking the materials to their own desks or tables. In either case, materials should be readily accessible, and students should be expected to take responsibility for cleaning up and returning materials to their appropriate storage locations. Giving a "5 minutes until cleanup" warning before the end of any session allows students to finish what they are working on and prepare for the upcoming transition.

Decide which arrangement to use in your classroom. You may need to experiment with a few different structures before finding the setup that works best for you and your students.

The Role of the Student

Establish clear guidelines when you introduce Choice Time. Discuss students' responsibilities:

- Try every choice at least once.
- Work with a partner or alone. (Some activities require that students work in pairs, while others can be done either alone or with a partner.)
- Keep track, on paper, of the choices you have worked on.
- Keep all your work in your math folder.
- Ask questions of other students when you don't understand or feel stuck. (Some teachers establish the rule, "Ask two other students before me," requiring students to check with two peers before coming to the teacher for help.)

For each Choice Time, list the activity choices on a chart, the board, or the overhead. Sketch a picture with each choice for students who may have difficulty reading the activity names. Some teachers laminate a piece of tagboard to create a

Choices board that they can easily update as new choices are added from session to session and old choices are no longer offered.

First grade students can keep track of the choices they have completed in one of these ways:

■ When they have completed an activity, students record its name or picture on a blank sheet of paper.

■ Post a sheet of lined paper at each station, or a sheet for each choice at the front of the room. At the top of each sheet, put the name of one activity and the corresponding picture. When students have completed an activity, they print their name on the corresponding sheet. Keep these lists throughout an investigation, as the same choices may be offered several times.

Some teachers keep a date stamp at each Choice Time station or at the front of the room, making it easy for students to record the date as well.

In any classroom there will be a range of how much work students complete. Some choices include extensions and additional problems for students who have completed their required work. Encourage students to return to choices they have done before, do another problem or two from the choice, or play a game again. You may also want to make the choices available at other times during the day.

Whenever students do any work on paper during Choice Time, they put this in their math folders at the end of the session.

At the end of a Choice Time session, spend a few minutes discussing with students what went smoothly, what sorts of issues arose and how they were resolved, and what students enjoyed or found difficult about Choice Time. Having students share the work they have been doing often sparks interest in an activity. Some days, you might ask two or three volunteers to talk about their work. On other days, you might pose a question that someone asked you during Choice Time, so that other students might respond to it. Encourage students to be involved in the process of finding solutions to problems that come up in the classroom. In doing so, they take some responsibility for their own behavior and become involved with establishing classroom policies.

The Role of the Teacher

Choice Time provides you with the opportunity to observe and listen to students while they work. At times, you may want to meet with individual students, pairs, or small groups who need help. This gives you the chance to focus on students you haven't had a chance to observe before, or to do individual assessments. Recording your observations of students will help keep you aware of how they are interacting with materials and solving problems. The **Teacher Note**, Keeping Track of Students' Work (p. 42), offers some strategies for recording and using your observations.

During the initial weeks of Choice Time, much of your time will probably be spent in classroom management, circulating around the room, helping students get settled into activities, and monitoring the process of moving from one choice to another. Once routines are familiar and well established, students will become more independent and responsible for their work during Choice Time. This will allow you to spend more concentrated periods of time observing the class as a whole or working with individuals and small groups.

Throughout the *Investigations* curriculum, there are numerous opportunities to observe students as they work. Teacher observations are an important part of ongoing assessment. A single observation is like a snapshot of a student's experience with a particular activity, but when considered over time, a collection of these snapshots provides an informative and detailed picture of a student. Such observations can be useful in documenting and assessing student's growth, as well as in planning curriculum. They offer important sources of information when preparing for parent conferences or writing student reports.

The way you observe students will vary throughout the year. At times you may be interested in particular problem-solving strategies that students are developing. Other times, you might want to observe how students use or do not use materials for solving problems. You may want to focus on how students interact when working in pairs or groups. Or you may be interested in noting the strategy that a student uses when playing a game during Choice Time. Class discussions also provide many opportunities to take note of student ideas and thinking.

You will probably need some sort of system to record and keep track of your observations. While a few ideas and suggestions are offered here, it's important to find a record-keeping system that works for you. All too often, keeping observation notes on a class of 28–32 students can quickly become overwhelming and time-consuming.

A class list of names is one convenient way of jotting down your observations. Since the space is somewhat limited, it is not possible to write lengthy notes; however, over time, these short observations provide important information.

Another common approach is to keep a supply of adhesive address labels on clipboards around the room. After taking notes on individual students, you can peel off each label and stick it in the appropriate student's file.

Some teachers keep a loose-leaf notebook with a page for each student. When something about a student's thinking strikes them as important, they jot down brief notes and the date.

You may find that writing notes at the end of each week works well for you. Some teachers find this a useful way of reflecting on individual students, on the curriculum, and on the class as a whole. Planning for the next week's activities often grows out of these weekly reflections.

In addition to your own notes, you will have each student's folder of work for the unit. This documentation of their experiences can help you keep track of your students, assess their growth over time, and communicate this information to others. An activity at the end of each unit, Choosing Student Work to Save, suggests particular pieces of work you might keep in a portfolio of work for the year.

Seven Peas and Carrots

What Happens

Students solve a problem in which they have seven peas and carrots altogether. They determine one or more combinations of peas and carrots they could have to make up seven in all. This is the first of several How Many of Each? problems they will solve throughout the school year. Student work focuses on:

- finding combinations of 7
- using manipulatives to help solve a problem
- using pictures, numbers, and words to record solutions to a problem
- finding more than one solution to a problem

Materials

- Counters (buttons, cubes, bread tabs) in at least two colors

Teacher Checkpoint

Seven Peas and Carrots

In this Checkpoint, you can get a sense of how your students think about number combinations, how they solve complex problems, and how they keep track of and record their work. See the **Teacher Note**, How Students Approach Seven Peas and Carrots (p. 50), for examples of how different students solve the problem.

The peas-and-carrots problem that students solve in this session is a type of problem we call How Many of Each? It is the first of several such problems they will solve throughout the school year. How Many of Each? problems always give the students two or more types of things (such as peas and carrots) and a total number (in this case, seven). The students determine how many of each type they could have to make up the total. These problems give students repeated practice with number combinations, with developing strategies for combining quantities, and with recording and organizing their solutions.

Introducing the Activity Bring a piece of chart paper and counters of two colors to the meeting area (ideally green and orange), and gather students around you. Draw a large circle on the chart paper to represent your plate.

I have seven things on my plate. Some of them are peas, and some of them are carrots. What could I have? How many peas? How many carrots? Remember, I have seven things in all. *[Write "7 in all" on the chart paper.]*

Accept two different student suggestions and model them with counters. If some students disagree with the solutions their classmates offer, ask them to explain their thinking, but keep the discussion brief, and keep the focus on explaining the task clearly enough so that students can find solutions on their own. Explain that there is more than one correct solution to this problem, and that students are to find solutions different from those just suggested.

Because students should be allowed to develop their own ways of recording their work for this activity, it's best *not* to record these first suggested solutions. Seeing the teacher model a recording method can make it difficult for students to find their own methods.

Solving the Problem Students may work alone, in pairs, or in small groups. Encourage them to share their ideas with one another. They may use counters to help solve the problem. If students are having difficulty beginning, encourage them to use concrete materials to model the problem. You might suggest they begin by taking a few counters of one color. Ask:

Suppose those are the peas. How could we find out how many carrots there are?

Or, you might put out a combination of eight or nine counters in two colors, and ask students if they can adjust the number of counters so there are seven in all. Be sure not to model a solution to the problem; students need the chance to find their own approaches.

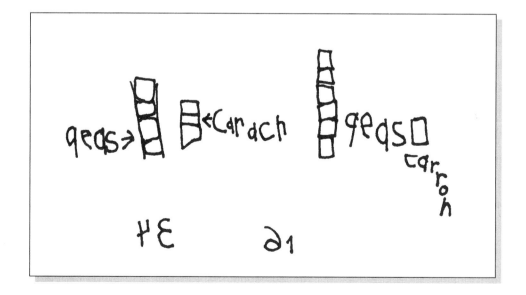

If students want to arrange counters on a "plate," they can draw a large circle on a sheet of paper. (Some teachers have used actual paper plates, but these tend to restrict the working space too much; students need plenty of room to arrange counters in different ways and move them around.)

When students have found a solution, they record it on blank paper using pictures, numbers, words, or a combination of these.

Observing the Students

As students are working, circulate to observe how they are approaching the problem and to offer support as needed.

■ What strategies do students have for approaching the problem?

Some students use strategies that involve combinations of seven, such as beginning with a number of counters less than seven and then counting up to seven to find how many more they need; starting with seven counters of one color and then replacing some with counters of a different color; or using number combinations that they know.

Other students solve the problem by collecting and counting a set of objects and adjusting the number until they have seven. For example, they count out seven counters, alternating one of each color and then counting the number of each. Or, they randomly take counters of two different colors, count them, and then add to or take away from the collection so that they have seven in all, perhaps recounting the set each time they make an adjustment.

■ How do students model the problem? Do they use manipulatives? pictures? numerals? Do they work mentally?

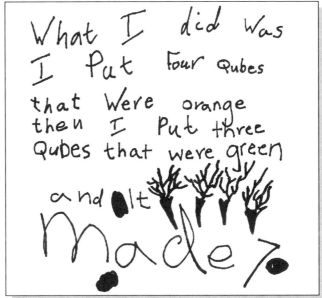

What I did was I put four Qubes that were orange then I put three Qubes that were green and it made 7.

- Can students count out seven things? Do they count from one each time, or do they count on from a number? Can they combine two sets and determine that there are seven in all? Can they keep track of the total number of peas and carrots and the number of each?

  If students are having difficulty working with combinations of seven, you might suggest they solve the problem with a smaller total, such as five. Keep track of any students you think need to work with a smaller number. In the next session, one of the Choice Time activities is a How Many of Each? problem with nine peas and carrots. Based on their work in this session, you may decide that some students should work with five or six peas and carrots instead.

- Do students check that their solutions are correct? What strategies do they have for checking?

- How do students record their solutions on paper? Do they draw pictures of peas and carrots? draw squares or other symbols to represent the counters they used to model the problem? write numbers? equations? words?

  All of these are important ways of showing mathematical thinking. Do not insist that students use one particular method, such as numbers, but do insist that they record in some way to show their solution. Students need many opportunities to discover what is important to record about a solution and to find different ways to show their work. For many students at the start of first grade, finding any way of recording solutions is challenging. Over time, as students share methods, they will develop more efficient ways of recording their work.

- Do students show how many peas and how many carrots (for example, "4 peas 3 carrots"), or do they just show how many in each group of things ("4, 3")?

  If some students record just the number in each group, ask them to find a way to show what each number represents. To point out the problem, you might ask them how you could tell whether they mean 4 peas and 3 carrots or 3 peas and 4 carrots. Some students may find it helpful to model the solution with counters of different colors before trying to show on paper what each number represents.

- Do students begin to look for more than one solution?

  One of the benefits of this activity is that all students can do it at levels that are appropriate and challenging for them. Some students may find just one solution; others may find pairs of "opposite" solutions (for example, 1 pea and 6 carrots; 6 peas and 1 carrot); others may find several solutions; and still others may challenge themselves to find all the possible solutions. For now, do not expect every student to find more than one solution. Throughout the year, as students work with a variety of problems that have many solutions, they will begin to seek multiple solutions on their own.

If Students Suggest Using Subtraction Some of your students may show ways to make 7 with subtraction, for example, with an expression such as 10 – 3. Acknowledge that 10 – 3 is 7, and then encourage them to think about whether subtraction yields an appropriate solution to the problem at hand. For example, if a student recorded one solution involving subtraction and another involving a combination of seven, you might say:

Your goal is to find a number of peas and a number of carrots that make up seven in all. You wrote down "2 p 5 c." Does that solve the problem? What does it show?

You also wrote down 10 – 3. Does that show the number of peas and the number of carrots you need to make seven in all? What does it show?

Students may recognize that 10 – 3 shows that if there are 10 things and 3 are taken away, there are 7 left. While it is important that they think about why this does not fit the peas and carrots problem, do not be concerned if some students still think they can solve the problem with subtraction. Over the year, they will work with a wide variety of addition and subtraction situations and explore relationships and differences between them. For now, students interested in using subtraction to make 7 can work on the extension, Ways to Make Seven (p. 48).

For Students Who Finish Early If some students find and record a solution before most of the others are finished, encourage them to share their work with a partner. If you think they are ready, they can look for additional solutions to the problem (see the **Teacher Note**, Exploring Multiple Solutions, p. 49). You might challenge some students to find all the solutions.

Some students will be eager to learn if their solution is correct. It's important to regularly ask students to explain their thinking behind an answer, whether it is correct or incorrect, and to encourage them to find ways to double-check their solutions on their own. See the **Teacher Note**, Encouraging Thinking and Reasoning (p. 97), for more suggestions on helping students to explain their thinking.

Activity

Sharing Solutions

About 20 minutes before the end of the session, call students together, bringing the papers on which they have recorded their solutions. Take about 10 minutes for a few students to share solutions, and leave the rest of the time for cleanup.

As a student holds up her paper and explains her solution, ask others to tell if they got the same solution.

Chanthou got 4 peas and 3 carrots. Did anyone else get that?

Record the solution on a piece of chart paper or on the board. Use one of the ways of recording that you observed students using as they worked on the problem. See the **Teacher Note**, When the Teacher Records Students' Solutions (p. 53), for further suggestions.

Did anyone get anything different?

Accept and record a few more solutions. After several minutes, you may find that some students are still eager to share and are attentive, but others are beginning to find it difficult to remain engaged and focused. Before ending the discussion, acknowledge that there may be still other solutions. You might provide an opportunity later in the day for those students still eager to continue sharing to do so with a partner or in a smaller group, or to individually add their solutions to your list.

Session 4 Follow-Up

 **Extensions**

Finding All the Solutions Students try to find all the solutions to the Seven Peas and Carrots problem, and they explain how they know they found all of them.

Peas, Carrots, and Blueberries This time, students have three different kinds of things: peas, carrots, and blueberries. They still have seven things in all. How many of each thing could they have?

Ways to Make Seven Students find and record different ways to make seven (such as 1 + 1 + 1 + 4 = 7 or 15 − 8 = 7). They may use calculators.

Exploring Multiple Solutions

Some of the problems in this unit (and through-out the *Investigations* curriculum) have more than one solution. Working on problems with multiple solutions opens up possibilities for clari-fying one's thinking and for exploring a variety of mathematical relationships: *Have I found all the solutions? Is there another way of thinking about the problems that would yield more solutions? Are there any relationships among the solutions I found?*

Exploring questions such as these can be very exciting for some students, but initially quite challenging for others. Many of us—at all levels of mathematical knowledge—are more familiar and comfortable with problems that have a sin-gle correct answer.

When students do the activity Seven Peas and Carrots (p. 43) at the start of first grade, they solve a problem that has many solutions. As you watch students working on this problem, you may notice a great deal of variation in their abil-ity to understand that some problems have more than one solution. Some may discover on their own that there is more than one solution. Others will realize this when they share their solutions with others. Still others may not be ready devel-opmentally to understand that some problems can have more than one solution. They may accept that students in the class have come up with different solutions, but they may not be ready to look for multiple solutions on their own. Or, they may believe that only one of the multiple solutions can be correct.

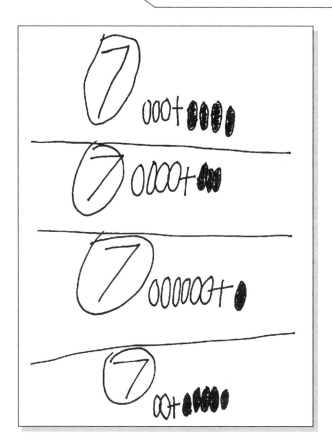

As your class works on problems with multiple solutions throughout the year, continue to pro-vide opportunities for students to share their solutions with others. As individual students become ready, encourage them to look for multi-ple solutions and to explore relationships among solutions (for example, "opposite" pairs of solu-tions, such as 6 peas 1 carrot and 6 carrots 1 pea, or sequences of solutions such as 1 pea 6 carrots, 2 peas 5 carrots, and 3 peas 4 carrots). You will find that over time, more and more of your students will begin seeking multiple solu-tions on their own and exploring relationships among them.

How Students Approach Seven Peas and Carrots

Seven Peas and Carrots is a complex problem that requires coordinating and keeping track of three pieces of information: the total number of peas and carrots, the number of peas, and the number of carrots. In order to solve the problem, students must count and keep track of a set of objects while comparing the number accumulated so far to the required total: *Do I have seven? Do I need more? fewer?* At the same time, they need to keep in mind how the two parts combine to reach that total: *If I take away two things so that I only have seven, now how many peas will I have? How many carrots? How can I change what I have to get a different combination?*

Students can solve a problem like this at levels that are appropriate and challenging for them. There is no single best way of approaching the problem or recording solutions, and you are likely to see many strategies and ways of recording work in your class. For example, one student might use combinations of seven that he knows to find his solutions, but choose to record his work with pictures of peas and carrots as a way of connecting the numbers and the things they represent. Another might use concrete materials to arrive at her solutions, but then record with numbers or equations. Another might use trial and error to find several solutions, notice some relationships among those solutions, and then use those relationships to find other solutions mentally. Yet another student may have a systematic strategy for using combinations of counters of two different colors to find all the possible solutions.

As you observe students at work on Seven Peas and Carrots, you can learn a lot about how they are thinking about number combinations, how they solve complex problems, and how they keep track of and record their work. The examples below illustrate the range of approaches observed in one class and the ways that the teacher supported students in working at an appropriate level and pace.

Thinking About Relationships Among Combinations

Soon after students begin working, the teacher visits Jacinta. She is counting silently on her fingers, working with intense concentration. After a moment, she relaxes and records:

$$2p + 5c = 7$$

She thinks for an instant, and then writes:

$$2c + 5p = 7$$

Only then does she look up at the teacher.

Can you tell me what you did to get your answers?

Jacinta: I figured out that two and five is seven. Then, I knew this one [2c + 5p] from this one [2p + 5c]. I was doing it backwards.

The teacher moves on and returns to Jacinta about ten minutes later. She has recorded:

$$2p + 5c = 7$$
$$2c + 5p = 7$$
$$3p + 4c = 7$$
$$3c + 4p = 7$$
$$2c + 1p = 7$$
$$2p + 1c = 7$$
$$7c + 0p = 7$$
$$7p + 0c = 7$$

Jacinta explains that she found ways to make seven, and then "made opposites." The teacher asks if she has found all the solutions, and Jacinta says she thinks so but isn't sure.

The teacher suggests that Jacinta compare her solutions with Eva, who has also found many solutions. She hopes that as they compare their work they will determine with more certainty whether they have a complete list and will perhaps begin thinking about how they know they have a complete list. They will also have an opportunity to share their different ways of organizing their solutions.

Difficulty Coordinating the Parts of the Problem

Garrett has just begun counting a set of green and orange counters when the teacher arrives. He tells her that he thinks there are seven. He counts out loud quickly, touching the counters as he counts. He skips two counters, and comes up with a total of nine. The teacher asks him to count again, slowly. He does this and gets 11, the actual number of counters. Again, she asks him to count slowly, and once more, he gets 11.

How can we fix that to make it seven?

Garrett: Take a little bit of this [green counters]. A little bit of this too [orange counters].

He takes off some of both color, without apparently counting the number he is removing. He seems to be focusing only on the fact that he has too many counters and that he needs to remove some. When he says he's finished, he has three orange and three green. He counts them and gets six.

How many do we want?

Garrett: Seven.

We have six here. What can we do to make it seven?

He adds one green counter and one orange counter, so there are now eight on his paper. The teacher asks him to count; he does, and gets eight. Again, she asks how many are needed in all, and he replies "Seven."

So, what could you do to get seven?

Garrett: I could take one out.

Garrett removes a counter and, on the teacher's prompting, recounts and gets seven. He is pleased, as he recognizes that he has found a solution. But he seems to think his solution consists of getting a total of seven things, rather than using a particular combination of peas and carrots to make up seven things.

How could we show how you solved the problem?

Garrett: We could make a seven. We did seven things.

That's right, we made seven things. How many carrots did we use?

Garrett: *[counting]* One, two, three.

How many peas?

Garrett: *[counting]* One, two, three, four.

So, we need to show how many peas and how many carrots you used to make seven in all.

Garrett decides to draw seven circles to show his solution. When he has finished, the teacher reminds him once again to think about how many of the seven represent carrots and how many represent peas. Garrett recounts the counters of each color on his paper and then colors in the circles accordingly.

Although Garrett is able to count up to seven objects accurately, he had difficulty coordinating the various components of the problem: counting and keeping track of a set of objects, keeping in mind the total number of objects needed, remembering to compare the number accumulated to the desired total, and keeping in mind the two parts he combined to reach that total.

The teacher decides that during Choice Time in Sessions 5 and 6, when students work on another How Many of Each? problem, she will ask Garrett to work with a total of four peas and carrots. With a smaller total, he may be better able to focus on the goal of the task and on relationships between the total and the two parts that make up the total. Once Garrett seems comfortable finding, explaining, and recording one or more solutions with a total of four, the teacher will ask him to work with a larger total, such as six.

The Beginnings of a Strategic Approach

Near the end of the session, the teacher visits Nadia. She has recorded the following:

Nadia explains that she is recording the combinations she is making with green and orange buttons. The lines on the left part show peas, and those on the right show carrots.

Are there more ways you can make seven peas and carrots?

Nadia: There might be. I could try five peas.

How did you think of five?

Nadia: Because it goes two, three, four, five.

She counts out five green buttons on her blank paper.

Nadia: I need to have seven, so I'll add two carrots. *[She picks up two orange buttons to place on her paper.]* One, that's six, and another one, that's seven.

She records her new solution:

Although Nadia appears to be working strategically and to have some sense of combinations of seven, she needs to work with concrete materials in order to find and verify her solutions. She is not comfortable recording solutions without first building them, as Jacinta did. It is not clear whether she is using relationships among the combinations that she has found to find new solutions; she may not notice these relationships, or she may simply not yet be able to articulate the relationships she does see. For example, she does not show that she recognizes that as the number of peas increase by one, the number of carrots decrease by one, or that she can use one solution to find "opposite" solutions.

The teacher encourages Nadia to continue trying to find all the solutions she can. She believes that as the girl continues her work on this problem and then solves other How Many of Each? problems, she will begin recognizing and articulating relationships among combinations and using these relationships to find new solutions.

When the Teacher Records Students' Solutions

When you are recording a solution for the class to consider, use one of the recording methods you saw students using: pictures, schematic representations (such as lines or dots), numbers, words. Varying the method will expose students to a range of representations and give them practice interpreting each.

For some discussions, you will want to focus the whole-group sharing on all the different ways students recorded their work (see the activity Sharing Recording Methods, p. 56). However, when you want to highlight the solutions, it's better to use just one method.

Students must understand that even though you are recording in a particular way today, they need not always record their solutions in this manner. Emphasize that you are using just one of the many valid recording strategies you saw them using in class:

All of you found good ways to record your work: with pictures of peas and carrots, with orange and green squares, with circles, with words and numbers. Today I'm using one of those ways to record your solutions on the board. Next time I'll use a different way to show your solutions, so we can see all the different ways you've found.

Plan to include numbers as you record, whether the students do or not. Recording three different solutions with pictures, your chart would look something like this:

Peas	Carrots
• •	🥕🥕🥕🥕🥕
2	5
• • • • • •	🥕
6	1
• • • • •	🥕🥕
5	2

On another day, you might record similar solutions in one of the following ways:

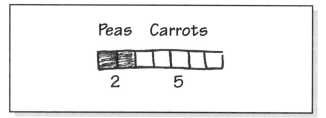

Record solutions in the order students give them, instead of organizing them in any way (for example, in order from smallest to largest number of peas, or linking "opposite" pairs, such as 5 peas 2 carrots and 2 peas 5 carrots). Later in the year this can change; you will occasionally want to model a particular way of organizing solutions, or a particular way of recording work (especially with numbers or with equations). However, in this first unit, avoid suggesting a particular way of recording. As students find their own ways of recording their work, they begin to think about what's important to communicate about a solution and what the different representations show. As they find their own ways to organize their work, they begin to seek out relationships among different solutions on their own.

If some students use equations, such as $3 + 4 = 7$, acknowledge this as a valid way of writing the relationship, but treat equations as just one of the many good ways of recording solutions. You don't need to model this method or require anyone to use it. Ask the student who uses it to explain what he or she means by it; you can tell students that they will be learning about equations later in the school year. (Equations are introduced in the grade 1 unit, *Building Number Sense.*)

Number Choices

Materials

- Number Cards, with wild cards removed
- Counters (buttons, cubes, bread tabs) in at least two colors
- Interlocking cubes (at least 30 per student)
- Staircase Cards 1–12 (and 13–20, optional)
- Staircase Sheet 3 (1 per student, homework)

What Happens

During Choice Time, students work on the following choices: solving a How Many of Each? problem involving nine peas and carrots, building staircases, and playing Compare or Double Compare. At the end of Session 6, they share ways that they recorded solutions to the How Many of Each? problem. Their work focuses on:

- combining two quantities
- comparing two numbers to find which is larger
- finding combinations of 9
- counting a set of objects
- using numbers to show how many
- ordering a set of numbers

Activity

Choice Time

Post a list of the choices and briefly remind students about their options. Explain that they will be working on these choices today and tomorrow. By the end of math time tomorrow, they must have done each of the choices. To help them remember what choices they have already completed, refer them to the records from the previous Choice Time (Sessions 2 and 3).

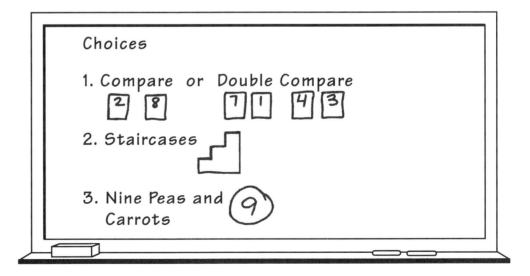

Note: If you think students no longer need to play Compare, remove it from the list.

To review the Choice 1 and 2 activities, see Compare (p. 26), Double Compare (p. 28), and Staircases (p. 34). If students who have been playing Compare seem ready for more challenge, introduce Double Compare to them.

Choice 3: Nine Peas and Carrots

Materials: Counters in two colors; paper

Students solve the following How Many of Each? problem:

> You have 9 things on your plate. Some are peas and some are carrots. How many of each could you have?

Record the problem somewhere for student reference (you might write it under the picture on your list of choices, or duplicate it on partial sheets of paper). This problem is identical to the one students solved in Session 4, except that here, they have *nine* peas and carrots. Students determine one or more combinations of peas and carrots they could have. Encourage them to share their ideas with one another. When they have found a solution, they record it on blank paper using pictures, numbers, words, or a combination of these.

If it seems necessary, remind students that they do not need to record their work in the way that you used when you recorded class solutions at the end of Session 4. They can use pictures, words, numbers, or a combination of these.

If you noticed some students having difficulty working with seven peas and carrots in Session 4, you might ask them to work with a total of five peas and carrots this time.

Observing the Students

Refer to page 38 for questions to consider while observing student work on Choices 1 and 2.

Nine Peas and Carrots

■ How do students approach the problem? Do they use counting? combinations of nine? Do students seem to work strategically, or do they use trial and error?

■ How do students model the problem? Do they use manipulatives? pictures? numerals? Do they work mentally?

■ Can students count out nine things? Do they count from one each time, or do they count on from a number? Can they combine two sets and determine that there are nine in all? Can they keep track of the total number of peas and carrots and the number of each?

■ Do students have strategies for checking that their solutions are correct?

- How do students record their solutions on paper? Do they show how many peas and how many carrots, or do they just show the number in each group?
- Do students recognize that the problem has more than one solution? Do they look for multiple solutions? Do they find relationships among the solutions?

Near the End of the Session At the end of each Choice Time session, remind students to record the choices they completed, either on their own lists or on lists you have posted. Also remind students to put their work in their folders.

Sharing Recording Methods

During the last 15 to 20 minutes of Session 6, call students together to share some of the ways they recorded their solutions to the Nine Peas and Carrots problem. (Students will need to bring their papers.) Take about 10 minutes for sharing solutions and leave the remaining time for cleanup.

I watched you recording your solutions to Nine Peas and Carrots in different ways. Some of you used pictures, some of you used numbers, and some of you used words. Who wants to share how they recorded?

As students share, do not discourage them from telling their solutions, but keep the focus on *how they recorded* their solutions. For emphasis, you might use chart paper to record a sample of different methods. This exposure to different recording methods, and the chance to explain their recording methods, is important for students.

Kristi Ann showed her solution with pictures of peas and carrots. Did anyone else show their solution with a picture? Did anyone show their solution a different way?

To get a good variety, call on students who have recorded in different ways. For example, look for a student who used numbers or equations, a student who used pictures *and* numbers, and a student who used words.

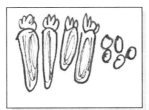 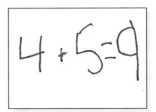 

Note: Consider whether your students would benefit from continuing with some of these Choice Time activities. The **Teacher Note**, Collaborating with the Authors (p. 58), talks about your role in making the *Investigations* curriculum meet the needs of your students. See also the **Teacher Note**, Games: The Importance of Playing More Than Once (p. 59), for a discussion of the value of repeating the games in the *Investigations* curriculum.

Sessions 5 and 6 Follow-Up

 **Homework**

Balls and Bats Student Sheet 3, Balls and Bats, offers another How Many of Each? problem. You might assign the same problem to everyone, or you might decide that different students work with different totals. In this case, you would modify the number on the sheet, for example, from 8 balls and bats to 6, or perhaps 13. Students find their own way to record their solution or solutions.

❖ **Tip for the Linguistically Diverse Classroom** To give students with limited proficiency in English a clear visual idea of what the two items in the problem are, show them an actual ball and bat or pictures.

Note: Consider sending home an example of the How Many of Each? problems students have already done in class, so that families can see the appropriate level of work and ways of recording solutions.

This curriculum is a guide, not a prescription or recipe. We tested these activities in many different classrooms, representing a range of students and teachers, and we revised our ideas constantly as we learned from students and teachers alike. *Every time* we tried a curriculum unit in a classroom, no matter how many times it had been tried and revised before, we discovered new ideas we wanted to add and changes we wanted to make. This process could be endless, but at some point we have to decide that the curriculum works well enough with a wide range of students.

We cannot anticipate the needs and strengths of your particular students this particular year. We believe that the only way for good curriculum to be used well is for teachers to participate in continually modifying it. Your role is to observe and listen carefully to your students, to try to understand how they are thinking, and to make decisions, based on your observations, about what they need next. Modifications to the curriculum that you will need to consider throughout the year include:

- changing the numbers in a problem to make the problem more accessible or more challenging for particular students

- repeating activities with which students need more experience

- engaging students in extensions and further questions

- rearranging pairs or small groups so that students learn from a variety of their peers

Your students can help you set the right level of pace and challenge. We have found that, when given choices of activities and problems, students often do choose the right level of difficulty for themselves. You can encourage students to do this by urging them to find problems that are "not too easy, not too hard, but just right." Help students understand that doing mathematics does not mean knowing the answer right away. Tell students often, "A good problem for you is a problem that makes you think hard and work hard. You might have to try more than one way of doing it before you figure it out."

The *Investigations* curriculum provides more than enough material for any student. Suggestions are included for extending activities, and some curriculum units contain optional sessions (called Excursions) to provide more opportunities to explore the big mathematical ideas of that unit. Many teachers also have favorite activities that they integrate into this curriculum. We encourage you to be an active partner with us in creating the way this curriculum can work best for your students.

Games: The Importance of Playing More Than Once

Games are used throughout the *Investigations* curriculum as a vehicle for engaging students in important mathematical ideas. The game format is one that most students enjoy, so the potential for repeated experiences with a concept or skill is great. Because most games involve at least two players, students are likely to learn strategies from each other whether they are playing cooperatively or competitively.

The more students play a mathematical game, the more opportunities they have to practice important skills and to think and reason mathematically. The first time or two that students play, they focus on learning the rules. Once they have mastered the rules, their interest turns to the mathematical content. For example, when students play Double Compare, they practice counting, combining, and comparing quantities. Over time, they become familiar with addition combinations through frequent experience, rather than by rote memorization.

For many students, repeated experiences with Double Compare lead quite naturally to developing more efficient strategies for combining numbers, to reasoning about numbers and number combinations, and to exploring relationships among number combinations. In other units, students will play games involving strategy. Once they are familiar with the rules, they can then begin to attend to strategic questions: *What's the best move I can make? How will my move affect the other player?*

Students need many opportunities to play mathematical games, not just during math time, but at other times as well: in the early morning as students arrive, during indoor recess, or as choices when other work is finished. Games played as homework can be a wonderful way of communicating with parents. Do not feel limited to those times when games are specifically suggested as homework in the curriculum; some teachers send home games even more frequently. One teacher made up "game packs" for loan, placing directions and needed materials in resealable plastic bags, and used these as homework assignments throughout the year. Students often checked out game packs to take home, even on days when homework was not assigned.

Patterns

What Happens

Session 1: What Comes Next? Students play a game in which they try to predict the color of the next cube in a cube pattern. They then make their own pattern sequences, using interlocking cubes.

Session 2: Clapping Patterns As a whole group, students act out patterns with gestures such as knees-knees-clap or clap-clap-knees-knees, then represent these patterns with colored cubes. They play What Comes Next? in pairs.

Sessions 3 and 4: Finding and Making Patterns Students compare the patterns they made in Session 1 and decide which are the same. During Choice Time, they play What Comes Next? with cubes, and make and record more of their own cube patterns. They identify patterns on their clothing and in the classroom and begin to make a pattern exhibit.

Sessions 5 and 6: What Is a Pattern? Students continue Choice Time with the addition of two more choices: playing What Comes Next? with pattern blocks, and making patterns for the pattern exhibit. They discuss their ideas about what a pattern is.

Routines Refer to the section About Classroom Routines (pp. 145–152) for suggestions on integrating into the school day regular practice of mathematical skills in counting, exploring data, and understanding time and changes.

Mathematical Emphasis

- Describing pattern sequences
- Predicting what comes next in a pattern sequence
- Constructing patterns from a variety of materials
- Classifying patterns as the same or different
- Distinguishing between things with and without patterns

What to Plan Ahead of Time

Materials

- Interlocking cubes: about 30 per student (Sessions 1–6)

- Paper cup (1 per student, to hold 3–4 cubes)

- Crayons, markers, or pencils in colors to match the interlocking cubes

- Scissors (Sessions 5–6, optional)

- Pattern blocks: 1 bucket for 6–8 students (Sessions 5–6)

- Geoblocks: class set (Sessions 5–6)

- Collage or construction materials for making patterns, such as paper or fabric scraps, buttons, beads, sequins, pasta shapes (Sessions 5–6)

Other Preparation

- Duplicate the following student sheets and teaching resources, located at the end of this unit. If you have Student Activity Booklets, copy only the items marked with an asterisk.

  For Session 1

  Cube pattern strips (p. 162): 1 sheet per student, cut into four separate strips

  For Sessions 3 and 4

  Student Sheet 4, Patterns from Home (p. 161): 1 per student, homework.

  For Sessions 5 and 6

  Pattern block cutouts* (pp. 163–168): 2–3 of each sheet on construction paper in the appropriate color (if you do not already have paper pattern blocks). Enlist adult help in cutting apart the shapes.

- Make an 11-inch paper tube, heavy or dark enough that the colors of a cube train won't show through. Cut a sheet in half the long way and roll a half-sheet around a train of 12 cubes. Tighten until the tube is just wider than the cube train. Cubes should slip through the tube easily, but remain well hidden. Tape or glue. (Session 1)

- Prepare three 12-cube trains, two in a-b-a-b patterns (but different colors), and one in either an a-b-c-a-b-c or an a-b-b-a-b-b pattern. Keep these hidden. (Session 1)

- Prepare a paper tube (as described above) for each student. While students *can* make these tubes, it is probably not worth the time; you can quite quickly prepare what you need for student use. (Sessions 2–6)

- Find objects, such as a checkerboard and a striped necktie, to use as examples for the pattern exhibit. Set aside a table or shelf in the room for this exhibit. (Sessions 3–4)

What Comes Next?

Materials

- Paper tube
- Three prepared 12-cube patterns
- Interlocking cubes (30 per student)
- Paper cups (1 per student)
- Cube pattern strips (1 or more per student)
- Crayons, markers, or pencils in colors to match the cubes

What Happens

Students play a game in which they try to predict the color of the next cube in a cube pattern. Then they make their own pattern sequences, using interlocking cubes. Their work focuses on:

- describing patterns
- predicting what comes next in a pattern sequence

Activity

Cube Patterns

Gather the whole class where they can talk with you and with each other. Each student needs three or four cubes, including one cube of every color you have used for your demonstration cube patterns. You'll need to establish rules about leaving their cubes alone while you are demonstrating the game; it helps if each student has the cubes in a small container, such as a paper cup. The cubes should be kept as individual cubes and not snapped together.

Show the tube with your first a-b-a-b cube pattern hidden inside (such as red-blue-red-blue-red-blue).

I'm going to show you a game called What Comes Next? I'm hiding a long row of 12 cubes in this tube. I've made it in a certain pattern. I'm going to show you a little bit of it. See if you can guess what color the next cube is—but don't tell anyone yet.

Here's what to do: If you think you can predict the color of the next cube, take a cube of that color from your cup and hide it in your lap. Don't say the color out loud. When everyone's ready, I'll ask you to show your cube. You don't have to show your prediction if you don't want to; you can just think about it in your head.

You may need to do this a couple times before students understand; just get started and they will catch on. First pull the cube train out of the tube so that only the first three cubes show.

Here are the first three cubes. What color are the first three cubes? [red, blue, red] If you think you know what color the next cube is, hide it in your lap. Shhh! Don't say the color out loud!

When you think everyone who wants to take a cube has done so, ask them to hold up their cubes. Take a quick look around to see what students are predicting, then have students put their cubes back in their cups. Show the next cube (the fourth cube) in your train.

Continue revealing one cube at a time and asking students to predict the next cube until you feel that most of them know the pattern. They may enjoy continuing to show cubes even when the pattern is well established.

Play again with another a-b-a-b-a-b cube train, in different colors. This time, let the students talk about their predictions.

Who would like to say what your prediction is? OK, Shavonne, you think green. Why do you think the next cube is green? Does anyone have something to add? Who thinks it might be a different color? Why do you think that?

Play the game a third time with the more complex pattern you have prepared, perhaps a-b-c-a-b-c (such as red, green, blue, red, green, blue) or a-b-b-a-b-b (such as red, green, green, red, green, green). This time, reveal the first two cubes. Most students will think that the next cube is red, as it would be if this were another a-b-a-b-a-b pattern. Have fun with this; you can tell students that this is a tricky one, that you've changed the pattern, that they'll have to think hard about this one. As you reveal more of the pattern, again ask students to share their ideas about their predictions.

Continue revealing cubes until you think most students have figured out the pattern or can't get any further in their reasoning. Then reveal the whole cube train.

What can you tell me about this pattern? Who has something else to say?

Throughout this game, try to emphasize the fun of making a prediction. Ask students for reasons, but don't expect all students to be able to make good predictions or to give reasons for their predictions. Encourage the idea that we can't always tell from the beginning of a pattern exactly how it will continue. See the **Dialogue Box**, What Comes Next? (p. 66), for a typical discussion in a first grade class.

Make Your Own Cube Pattern

To prepare for playing What Comes Next? with a partner, students make cube trains in patterns of their choice. For this activity, students work together at tables or groups of desks. Each table needs a supply of interlocking cubes (about 30 per student).

Each student makes two different patterns, using 12 cubes for each pattern. They record their favorite on one of the cube pattern strips you have prepared.

Note: As students develop their patterns, it's likely that they will need more cubes of a certain color, or want to use a color not available at their table. Discuss some guidelines for sharing: How would they ask another student or table for cubes they need? How would they respond to someone's request for cubes? Students might role-play this situation for practice.

As you circulate and watch students work, ask them about their patterns.

Tell me about your patterns. How are your two patterns different? If I covered up this much of your cube train, how would you know what color cube comes next? Do you see any patterns at this table that are the same?

At this point, you don't need to define the word *pattern*; just continue to use it in context as you ask students to describe their patterns. In Session 6, you will ask students for their ideas about what a pattern is. Don't worry at this point if some students don't seem to distinguish between a pattern and any random color sequence. Simply take note of students' different ideas as they talk about their patterns.

Once again you are likely to see a wide range of understanding in your class. Many students will be able to think about an a-b-a-b-a-b pattern easily, but may have more difficulty following or constructing a pattern with three parts in each pattern unit (such as a-a-b-a-a-b, or a-b-c-a-b-c). Some may not yet have the idea of a pattern that can be predicted. Others may be able to develop quite complicated patterns, and even series of patterns that are related to each other.

As students see many patterns, construct their own, and discuss various types, they will deepen their understanding of what a pattern is and how it can be described. For some background on the mathematical issues in constructing or recognizing a pattern, refer to the **Teacher Note**, Seeing Patterns (p. 65).

Collect students' patterns or have them put them in their notebooks or folders. They will need them for discussion at the beginning of Session 3.

Seeing Patterns

Mathematics has been called "the science of patterns," for it is often used as a language to describe and predict numerical or geometric regularities. Some of these patterns are very simple. When you count by 2's on a 100 chart, you and your students can see patterns in the way the numbers are arranged on the chart. If you list these multiples of 2 (2, 4, 6, 8, 10, and so forth), you can begin to see patterns in the numbers themselves. These patterns are not accidental; they indicate important aspects of our number system and of mathematical relationships. As students become aware that patterns exist in mathematics, they start seeing them everywhere and using patterns to make predictions.

In this investigation, students are learning to think about regularity and repetition. A pattern sequence, like the sequences students make with the interlocking cubes, is made up of repeating units. In an a-b-a-b-a-b pattern, the unit is a-b. Repetitions of this unit, one after the other, make the pattern. As units get larger and more complex, it is sometimes difficult to look at a pattern and figure out the basic unit. The unit for the pattern a-b-b-a-b-b-a-b-b is a three-element unit, a-b-b. However, the unit for the pattern a-b-b-a-a-b-b-a-a-b-b-a is harder to see. (This pattern shows three repetitions of the basic unit a-b-b-a.)

Many of your students will not think of a pattern as made up of units. Rather, they will think in terms of "what comes after what." For example, for the cube sequence yellow-red-yellow-red-yellow-red, students won't necessarily think about the pattern as being made from the unit yellow-red. Rather, they will think of the pattern as "red comes after yellow, yellow comes after red." They think about the action of building a sequence in order to understand the pattern. You will notice some differences in the students' descriptions, depending on whether they are beginning to see the units in a pattern or are following the sequence, cube by cube, as in the **Dialogue Box**, What Comes Next? (p. 66).

Not all patterns are linear sequences, like the cube patterns. Think about a quilt pattern or a pattern your students may have made with pattern blocks. These patterns might be built out from the center, or they may repeat in more than one direction. You have probably seen patterns on wallpaper, wrapping paper, or floor tiles that you were convinced were built in some systematic way, but were quite difficult to figure out. To become more aware of the variety and complexity of pattern, look at some patterns yourself and see if you can describe where the pattern begins and how it repeats. Can you predict what comes next in any direction?

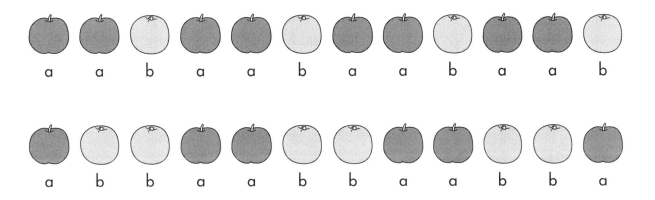

What Comes Next?

Students in this class are learning how to play What Comes Next? with their teacher (p. 62). Each student has a paper cup with one green cube, one white cube, and one orange cube. The class has already worked with a-b-a-b-a-b patterns, and this time the teacher has hidden an a-b-c-a-b-c pattern in the tube: white-green-orange-white-green-orange. The first four cubes are showing.

So we have white-green-orange-white so far. Hold up the cube you think comes next. *[Pause.]* **Yanni, why did you pick green?**

Yanni: Because a pattern keeps repeating.

Tamika: It's got another pattern. White, green, orange and it keeps on repeating itself over and over.

Anyone have something to add? Andre, you picked green, too. Why?

Andre: Because the green comes after the white.

Libby: Because the pattern goes white, green, orange, white, green, orange.

OK. Here's a new pattern. I'm not going to start until all the cubes are back in the cups.

The teacher shows two cubes, green followed by orange. There are a lot of different guesses of what might come next; all three colors are named as possibilities. The teacher reveals that the third cube is orange; the cubes showing now are green-orange-orange. Most students predict a green cube next, although Fernando shows a white one.

Who wants to tell us a reason for the color they're holding up?

Claire: I think green goes next. The pattern goes green, orange, orange, green, orange, orange.

Kaneisha: I picked green because I knew it was green at the top and then orange orange, so next would have to be green.

Why did you choose white, Fernando?

Fernando: Well, it's probably green, like Kaneisha said, but it could be white. I was thinking green-orange-orange and then two whites and then it would start again.

So Fernando is saying it's possible that we may not have seen the whole pattern yet. It could go green-orange-orange-white-white, then green-orange-orange-white-white again. So, let's see one more cube, and then see what everyone thinks.

Most of these students seem to be able to think about a pattern made from a three-element unit. Fernando is thinking of even more possibilities. He seems to have the idea that the unit of a pattern can be made out of even more than three cubes, as he suggests a possible five-element unit. The teacher knows that not all students are following Fernando's explanation, but restates his idea for those students who might begin to consider it.

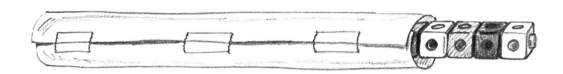

Clapping Patterns

What Happens

As a whole group, students act out patterns with gestures such as knees-knees-clap or clap-clap-knees-knees, then represent these patterns with colored cubes. They play What Comes Next? in pairs. Their work focuses on:

- representing patterns in more than one way
- creating patterns with cubes
- predicting what comes next in a pattern

Materials

- Interlocking cubes (30 per student)
- Paper tubes (1 per student)

Clapping Patterns

Gather students where they can easily see you as you sit in a chair or on the floor (a circle works best) to introduce the clapping patterns.

Begin a clapping pattern and ask students to join in, doing it with you. Start with a knees-knees-clap pattern: hit your knees with your hands twice, then clap. This should have a steady rhythm to it, one beat for each knee-slap or clap.

Try one or two more patterns. You might do, for example: knees-clap-clap and knees-knees-clap-clap.

Return to the first pattern (knees-knees-clap). Do it again with the students and ask them to describe it to you in words.

What could we say to describe this pattern? Who can tell me something about it? Who can say something else?

From a container of cubes, pull out a few cubes in two colors. Ask students to help you model the knees-knees-clap pattern with cubes:

Let's use red for knees and blue for clap. How could I make a cube tower to show our knees-knees-clap pattern?

If students don't understand this question, do the clapping pattern with them again. This time, while you are clapping, say the words "red, red, blue, red, red, blue" as you clap. Then ask your question again.

Next try working in the other direction, from cubes to gestures. Using one of the patterns a student made in Session 1 or a simple cube pattern you have made, ask students to figure out a way that the class could clap out the pattern. For example, if the cube pattern is red-blue-blue-blue-red-blue-blue-blue, students might decide on a clapping pattern of knees-clap-clap-clap-knees-clap-clap-clap.

You may want to try a pattern that requires a third gesture, such as putting both hands on your head. For example, the cube pattern red-blue-green-red-blue-green could be done with the gestures knees-clap-head-knees-clap-head.

As students work on cube patterns during the rest of this investigation, translating them into gestures may help some students think about their patterns. You can continue to do this activity when you have a spare few minutes outside of math time as well.

Activity

What Comes Next?

Remind students of the game you played yesterday with the cubes in a tube, when they guessed what cube came next in your pattern. Today they will play this game in pairs. Distribute the paper tubes, one to each student. Each student makes a pattern with 12 cubes and hides the train in the tube. Partners then take turns revealing their train one cube at a time, while the other student predicts the color of the next cube.

To help students keep their pattern secret from their partner, each person in a pair could work at a different table until they are both ready to play.

Students will have more time to play this game during Choice Time in the next few sessions.

Session 2 Follow-Up

 **Homework**

Double Compare Students play Double Compare at home for continued practice with number combinations. Check to be sure that everyone has taken home a copy of the appropriate game directions. They should still have their home set of Number Cards from the previous investigation.

Finding and Making Patterns

What Happens

Students compare the patterns they made in Session 1 and decide which are the same. During Choice Time, they play What Comes Next? with cubes, and make and record more of their own cube patterns. They identify patterns on their clothing and in the classroom and begin to make a pattern exhibit. Their work focuses on:

- comparing patterns
- predicting what comes next in a pattern
- creating patterns with cubes
- finding patterns in the world around them

Materials

- Interlocking cubes (30 per student)
- Students' paper tubes
- Cube pattern strips (several per per student)
- Crayons, markers, or pencils in colors to match the cubes
- Student Sheet 4 (1 per student, homework)

Activity

Gather the whole class in a circle, asking students to bring with them their record of a cube pattern they made in Session 1 (the cube pattern strip they colored).

Who has a pattern they'd like to show? OK, Jonah, tell us about yours.

After one student has shown a pattern, ask if anyone has a pattern that's the same.

Who has a pattern that you think is the same as Jonah's?

Students who have a pattern they think is "the same" can hold it up. Or, these students could stand up in a row and hold their patterns in front of them.

Who can tell us something about what makes these patterns the same?

Accept students' comments without making any judgments or saying that you agree or disagree. If students have different opinions, you can acknowledge this.

Same or Different?

Susanna thinks Leah's pattern is the same as Donte's because they both have two colors and they're both made by alternating one cube of each color. This one's red, green, red, green. This one's yellow, blue, yellow, blue. But Chris thinks the patterns are different because they have different colors.

After students have commented on the first set of patterns, collect all the patterns in this group. Clip them together, because you'll be posting them later as a set.

Now, does anyone have a different pattern to show?

After seeing another student's pattern, again ask any students who have a pattern they think is the same to show theirs. Ask for student comments, then collect this group of patterns as well. Ask if anyone has another pattern that is different from those shown so far. Without making this meeting too long, you will probably be able to show every child's pattern since most are likely to fall into two or three groups. Collect and clip each group as a separate set.

If some students have patterns that you wouldn't consider a pattern, accept student comments about this without making a judgment.

Diego thinks that this one isn't a pattern because he says, "The colors aren't in an order. They just go any way." This is an interesting question for us to think about. We're going to talk more about what a pattern is in the next few days. Can anything that has two colors like this be a pattern? Is any picture or design a pattern? Keep thinking about this, and I'll ask you again about your ideas.

Note: Sometime before Session 4, perhaps once the students have started the Choice Time activities, post these patterns in their groups. You might designate each one with a number to facilitate later discussion: Pattern 1, Pattern 2, and so forth.

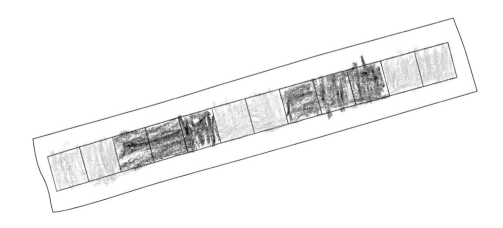

Choice Time

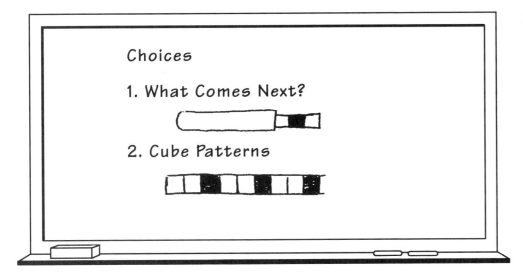

Post two choices for Sessions 3 and 4, including a picture with each.

Choice 1: What Comes Next?

Materials: Interlocking cubes, paper tubes

Students continue playing the game they started in Session 2. Once both students in a pair have had their pattern guessed, they might each make a new 12-cube pattern and play again, or they might switch partners, using the same cube pattern again.

Choice 2: Cube Patterns

Materials: Interlocking cubes; cube pattern strips; crayons or markers.

Students make more cube patterns like the ones they made in Session 1. They may try several different patterns, but eventually should choose one to record on a cube pattern strip for the class display. They could make one that fits one of the groups that have already been identified, or they might make one that they think is different from any patterns you have so far. If you have time while students are working, post the cube patterns, grouped by type; then students can check this display to see whether their pattern is a new one or not.

As students continue their work on this activity during the rest of this investigation, keep adding their finished patterns to the class display. Ask students to help you decide whether their pattern fits a group that is already posted or belongs in a new group.

You will probably have some patterns that clearly "go together" and others that are unique. Watch for patterns that bring up interesting issues for the whole class to discuss at the end of Session 4 (Where Does This Pattern Fit? p. 73). Some students may need to extend their patterns beyond the 12-cube limit in order to carry out their ideas. See the **Teacher Note**, First Graders' Cube Patterns (p. 75), for examples of the variety of patterns students make.

If some students seem confident in making patterns, you might ask questions about the *number* of cubes they are using. For example, if a student is making a red-blue-red-blue pattern, you might ask:

When you have all 12 cubes in your cube pattern, how many red cubes will there be? Go ahead and make your pattern and find out.

So you have six red cubes. How many blue cubes do you think there are? What if you made your pattern with ten cubes, then how many red cubes do you think there would be?

Observing the Students

During Choice Time, observe students to get an overall sense of what they think a pattern is.

What Comes Next?

- Do students use the cubes that are showing to help predict the next cube, or are they just guessing?
- Once a few units of the pattern are showing, do students see the sequence?
- Can some students predict a-b-a-b-a-b patterns, but not other patterns?

Making Cube Patterns

- Can students record their cube pattern accurately?
- Can they recognize when their cube pattern is the same as or different from other patterns?
- Do students make a variety of patterns? Do some students only make a-b-a-b-a-b patterns? Do some students make patterns with a basic unit of three cubes (such as a-b-c-a-b-c, or a-b-b-a-b-b)?
- Do some students make random color sequences that don't appear to have a pattern?

Patterns Around Us

During the last 15 minutes of Session 3, after a quick cleanup, gather students for a discussion about patterns you can see in the classroom:

I'd like Yukiko to stand up. I asked her to stand up because she's wearing a shirt that has a pattern. Why do you think I called this a pattern? Who sees someone else who is wearing a pattern?

Tuan, you said Brady is wearing something with a pattern. How does Brady's pattern go?

Ask students to look around the classroom and point out anything else they see that has a pattern and describe any pattern they find. Keep in mind that the class will further discuss what a pattern is in Session 5, after some more pattern experiences.

Can you think of anything at home that has a pattern?

Students might name things like wallpaper, floor tiles, dishes or glasses, pieces of clothing, wrapping paper, towels. After hearing a few suggestions, explain that over the next few days, you'll be making a pattern exhibit. Students should bring any small objects from home that they are allowed to borrow for the exhibit (ask them not to bring anything breakable). Student Sheet 4, Patterns from Home, can be sent home to alert families to this assignment. Students who have trouble finding something to bring can make patterns for the exhibit, using materials they have at home.

Take a little time to talk about the exhibit. Show the class where it will be set up. Explain that students will need to take special care of the things in the exhibit, just like a museum exhibit. If you have brought a couple of items to start the exhibit, show them to the class and place them on the exhibit area. Some teachers have draped the area set aside for the exhibit with a piece of plain cloth as a background for all the patterns.

Where Does This Pattern Fit?

Take a few minutes at the beginning or end of Session 4 to talk about one or two of the patterns students have been coloring and adding to the class display. For examples of typical patterns you might want to discuss, see the **Teacher Note**, First Graders' Cube Patterns (p. 75).

If there are any disagreements about where a pattern fits on the display, be sure to talk about these. For example, one student made a pattern like this:

green red green red green red red green red green red green

Some students said that this pattern was "wrong," and that "there's a mistake in the middle." Other students thought it belonged with the a-b-a-b-a-b pattern group, while still other students thought it was a different kind of pattern. The teacher thought this was a good pattern for the whole class to discuss because she noticed that the pattern was symmetrical, and she wondered if any of her students would be able to talk about this symmetry in their own words.

Sessions 3 and 4 Follow-Up

Patterns from Home After Session 3, send home Student Sheet 4, Patterns from Home, to tell families about your pattern exhibit. Remind students to look for things at home that they could borrow for the exhibit, or to make patterns of their own using materials at home.

First Graders' Cube Patterns

These examples of the kinds of patterns first graders have made with interlocking cubes have been selected to illustrate how complex and multifaceted the idea of *pattern* is, how inventive students are as they construct their own patterns, and what difficulties some students encounter as they build patterns.

Many first graders start out with cube patterns of two colors in an a-b-a-b pattern. As they gain more experience and see other examples, some begin to build patterns with three parts in each unit, such as a-b-b-a-b-b or a-b-c-a-b-c.

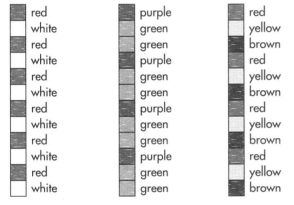

As students experiment more, you will start to see a wide variety of ideas. Some students may attach 12 cubes without any repetition or regularity that you can see (or that they can describe); they may not yet understand what a pattern is.

Other students may begin to repeat some elements, but will not develop a pattern that can be completely predicted. For example, you might see some cube trains like the following:

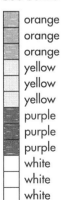

This student is thinking of the sequence as three of one color, followed by three of another color, followed by three of another color. He has the idea of a repeating element—everything comes in threes—but he hasn't yet coordinated the repetition of both number and color to make a pattern with a predictable repetition.

You may also see this kind of pattern. There is certainly some regularity about this pattern, but it isn't completely predictable. There is no way to know what color comes after black, although there is a way to predict what comes after any cube that is not black.

These students are working on the idea of a repeating pattern. As they explore their own ideas and see the ideas of other students, as well as patterns that you construct, they will gradually think more about how to construct a pattern that can be described and predicted.

One student made a series of related patterns:

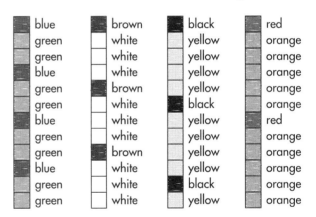

The teacher used this student's ideas as a focus for discussion, asking her classmates to describe what they noticed about her cube trains. This discussion allowed a range of students to make observations. Some simply described one pattern: "The first one goes blue then green then green." Another related the color pattern to a number pattern: "She went in order—2, then 3, then up to 5." Some made predictions about what the next cube train would look like.

Some students will want to make patterns that can't be contained in a 12-cube sequence. They may have in mind a longer pattern, or several patterns that relate to each other. For example, one student made the following pattern:

■	black
■	black
■	black
▨	red
▨	red
▨	red
□	white
□	white
□	white
■	black
■	black
▨	red
▨	red
□	white
□	white
■	black
▨	red
□	white

The class talked about how this pattern worked and observed a pattern in the number of cubes as well as in the sequence of colors. This pattern is an interesting one because it seems finished. It goes 3, 2, 1, and because we can't have fewer than one block, the sequence is over. However, the teacher asked students what they might do to extend the pattern.

Several suggestions were made, including starting over again with 3 blacks, or building up again from 1 to 2 to 3 of each color. Together, they built a new sequence that looked like this:

■	black
■	black
■	black
▨	red
▨	red
▨	red
□	white
□	white
□	white
■	black
■	black
▨	red
▨	red
□	white
□	white
■	black
▨	red
□	white
■	black
■	black
▨	red
▨	red
□	white
□	white
■	black
■	black
■	black
▨	red
▨	red
▨	red
□	white
□	white
□	white

Looking at and describing each other's patterns helps all students expand their ideas of what a pattern is.

What Is a Pattern?

What Happens

Students continue Choice Time with the addition of two more choices: playing What Comes Next? with pattern blocks, and making patterns for the pattern exhibit. They discuss their ideas about what a pattern is. Their work focuses on:

- predicting what comes next in a pattern sequence
- finding and creating patterns
- describing what a pattern is

Materials

- Interlocking cubes (30 per student)
- Pattern blocks (1 bucket per 6–8 students)
- Paper tubes (1 per student)
- Cube pattern strips (several per student)
- Crayons, markers, or pencils in colors to match the cubes
- Scissors (optional)
- Pattern block cutouts
- Geoblocks
- Colored paper or fabric scraps, buttons, beads, sequins, pasta shapes (optional)

Activity

Starting the Pattern Exhibit

As students bring in objects from home for the pattern exhibit, help arrange them carefully on the exhibit table. Students may want to make display signs on index cards, naming the object and who brought it.

Throughout the day, encourage students to take a look at the objects in the exhibit. If you have a regular sharing time in your class, you can use this time for a few students to share the exhibit items they brought.

Activity

Choice Time

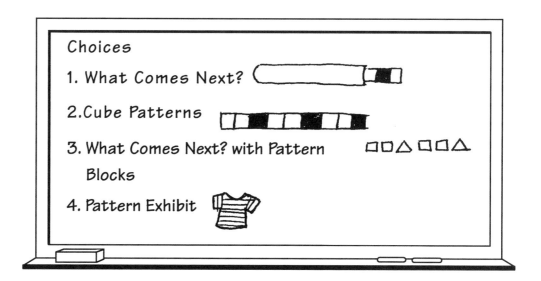

Post four choices, adding two new activities to the previous Choice Time list.

Some students may want to return to the first two choices, but all students should try Choice 3, What Comes Next? with Pattern Blocks, sometime during this Choice Time. If most students have completed both of the first two choices earlier, most of the class may be working on Choices 3 and 4 in these two sessions.

Spend a few minutes introducing the two new choices. Show the class how to play What Comes Next? with Pattern Blocks, and explain the Pattern Exhibit choice.

For a review of the descriptions of Choices 1 and 2, see pp. 71–72.

Choice 3: What Comes Next? with Pattern Blocks

Materials: Pattern blocks, sheets of paper or cardboard to cover the pattern block sequence

This version of What Comes Next? is similar to the version with cubes. Working at separate tables, each of a student pair makes a pattern of 12 pattern blocks laid out in a single row. For example:

To introduce What Comes Next? with Pattern Blocks, play a demonstration round for the whole class. For this demonstration, emphasize making a pattern in one long line, as the fit of pattern blocks tempts students to create radiating patterns that are not suitable for What Comes Next?

When the sequence is finished, the student covers it with a sheet of paper or cardboard (or a book), leaving only the first two or three blocks showing. Students then meet with their partners and take turns trying to predict the next block in the sequence. The block is revealed by sliding the paper or book so that it shows the next block. The student continues to guess each successive block until the entire pattern is revealed.

Choice 4: Pattern Exhibit

Materials: Paper, crayons; pattern blocks and Geoblocks; pattern block stickers or cutouts (optional); collage or construction materials for making patterns, such as fabric scraps, buttons, beads, sequins, pasta shapes, colored paper scraps (optional); scissors, glue (optional)

Students make a pattern for the class pattern exhibit. They might draw patterns with crayons, cut and paste paper shapes in different colors, or assemble materials in patterns and glue them to heavy paper. If you have pattern block stickers or cutouts, students can make and record pattern block designs.

Ask students to make drafts of their work before doing a final piece for the exhibit. If they are drawing, they can make a sketch with pencil before using crayon; if they plan to cut out and glue things, they should assemble all the pieces for their pattern before they glue them down. This "first draft" process encourages students to think through their patterns before they begin to color or glue.

This activity gives you a chance to see how students think about pattern when they make up their own. Both the cubes and pattern blocks provide their own structure; certain shapes and colors are built into these materials. The way they fit together also naturally suggests certain ways to make patterns. When students make their own patterns, you will probably see much more variety, and thereby gain more insight into their thinking about pattern. Also, since students are not constrained by the materials, you may see more examples that, to you, do not seem to be patterns. Avoid passing judgment about whether something is or is not a pattern; instead, ask students to describe their patterns to you and to other students.

As students finish, you may want to mount their patterns on construction paper and hang them behind the exhibit table.

Observing the Students

As students work on the choices, continue to observe their developing sense of pattern.

- Are they making a variety of patterns (not just a-b-a-b, but also a-b-c-a-b-c, a-a-b-a-a-b, a-b-b-a-b-b, and others)?
- Can they predict a sequence from seeing a few units of the pattern? Can they predict some patterns but not others?
- Can they record their patterns accurately?
- Can they recognize patterns that are the same and patterns that are different?

Activity

What's a Pattern?

At the end of Session 6, gather students for a whole-class discussion of pattern. Be sure that students have had a chance to look at the pattern exhibit and discover the patterns both in objects they have brought from home and the patterns they have made themselves. Start the discussion by holding up a few objects from the exhibit, one at a time.

Is this a pattern? How do you know? How does this pattern go? Is there a way to tell what comes next at different places on the pattern?

Can you point to something in the classroom that doesn't have a pattern? Why do you say it doesn't have a pattern?

Write down a few of the things that students say about patterns on chart paper and post them near the pattern exhibit.

❖ **Tip for the Linguistically Diverse Classroom** Include a simple sketch next to each written item so that nonreaders will be able to understand the chart.

You may want to leave the exhibit up for another week or so and have students continue to add to it.

INVESTIGATION 4

Counting and Combining

What Happens

Session 1: Collect 15 Together Students play Collect 15 Together, a cooperative game in which they work together to collect at least 15 counters. As they accumulate counters, they must keep track of the number they have in all.

Sessions 2 and 3: Counting and Combining During Choice Time, students work on three activities: Collect 15 Together (or a more advanced variation of the game); What Comes Next? with their choice of materials; and either of the card games learned earlier, Compare or Double Compare.

Session 4: Eleven Fruits As an assessment, students solve a How Many of Each? problem about eleven fruits (or another number, as needed to adapt the level of difficulty to your class). Students then share their solutions and solution strategies in a class discussion.

Session 5: Making Predictions About a Story Students listen to the first half of a story in which a total grows by accumulating progressively larger groups (such as *Rooster's Off to See the World*). They use pictures, numbers, and words to show what they think happens next in the story. A few share their ideas with the class.

Session 6: How Many in All? Students listen to the rest of the story that was started in Session 5 and determine how many in all are in the story. They record and share their work.

Routines Refer to the section About Classroom Routines (pp. 145–152) for suggestions on integrating into the school day regular practice of mathematical skills in counting, exploring data, and understanding time and changes.

Mathematical Emphasis

- Counting and keeping track of a set of objects
- Extending strategies for comparing two quantities
- Using counting, patterns, and other strategies to help solve problems
- Extending understanding of number combinations
- Extending strategies for combining two quantities
- Representing solutions to problems with pictures, numbers, and words
- Making and explaining predictions

What to Plan Ahead of Time

Materials

- Chart paper or newsprint (18 by 24 inches): 15–20 sheets (available for teacher use as needed)

- Blank paper (available for student use as needed)

- Dot cubes: 1 per pair (Sessions 1–3)

- Pattern blocks: 1 bucket for each 6–8 students (Sessions 2–3)

- Geoblocks: class set (Sessions 2–3)

- Interlocking cubes (available for student use as needed)

- Number Cards (Sessions 2–3)

- Counters, such as buttons, bread tabs, or pennies: at least 30 per student (available for student use as needed)

- *Rooster's Off to See the World,* by Eric Carle (Picture Book Studio, 1987) or a similar story (Sessions 5 and 6, optional; see note)

Note: In Sessions 5 and 6, you read a book or tell a story in which first there is one thing (or animal or person), then two more join them, then three join them, and so on. The activities described in Sessions 5 and 6 are based on *Rooster's Off to See the World,* but you can substitute any children's book with a similar theme, such as *The Very Hungry Caterpillar,* also by Eric Carle, or *P. Bear's New Year's Party* by Paul Owen Lewis. If you cannot get one of these or another book that illustrates this pattern, suggestions for making up your own story are given in Session 5.

Other Preparation

Duplicate the following student sheets, located at the end of this unit. If you have Student Activity Booklets, copy only those items marked with an asterisk.

For Session 1

Student Sheet 5, Collect 15 Together: 1 per student (homework)

For Sessions 2 and 3

Game Record Sheet* (p. 178): 1 per student (homework, optional)

Student Sheet 6, Dinosaurs and Tigers (p. 170): 1 per student (homework). Consider modifying the number on this sheet for some students.

Collect 15 Together

Materials

- Dot cubes (1 per pair)
- Counters such as buttons, bread tabs (about 25 per pair)
- Student Sheet 5 (1 per student, homework)

What Happens

Students play Collect 15 Together, a cooperative game in which they work together to collect at least 15 counters. As they accumulate counters, they must keep track of the number they have in all. Their work focuses on:

- counting a group of objects
- keeping track of the size of a growing collection of objects

---

Activity

Collect 15 Together

Introduce Collect 15 Together to the entire class by assembling students in a circle on the floor and playing a demonstration game. Either enlist two student volunteers to play, or play the game yourself with a student partner.

Today we're going to play a game called Collect 15 Together. You will play with a partner. The two of you will work together, rolling a dot cube, until you collect 15 counters. Nobody loses; you both win when you can show that you've collected at least 15.

Players take turns. The first player rolls a dot cube and takes that many counters. Let's watch these two boys play, to see how it goes.

Jamaar rolled 2. How many counters does he put in a group?

Max rolled 5. How many counters does he add to the group?

Do you think they have 15 counters yet? How many do they have? Is that more or less than 15?

Ask two or three students to explain how they figured out how many counters there are in all. Some students may use counting strategies, such as counting on or counting all the counters. Others may use knowledge of number combinations.

Play for another turn or two, or until you think students understand the game. Continue to ask students to explain how they figured out how many counters there are in all.

Explain that the game is over as soon as the players have at least 15 counters. Depending on what they roll, they may get 15 exactly, or they may end up with a few more than 15.

After your demonstration, pair students to begin play. Each pair needs a dot cube and about 25 counters.

Observing the Students

Circulate to observe students playing the game and to offer support as needed.

- Do students understand the rules of the game?

- How comfortable and accurate are students as they count the counters? What sorts of errors do you notice in their counting? Are they able to keep track of what has been counted and what needs to be counted?

- How do they find the total number of counters after the new ones have been added at each turn? For example, do they count all the counters? count on from the number left after the last turn? use mental computation? count on their fingers? organize the counters into groups and count the groups?

- Do they have a way of keeping track of the number of counters at the end of each turn? For example, do they commit the number of counters to memory? organize the counters in some way? write the number each time?

- Do they recognize when they have reached 15? Do they recognize when they have reached a number greater than 15?

- Do students play cooperatively and help one another?

If some students are having difficulty, suggest they play to collect a smaller number, such as 10. As you see students who seem ready for more challenge, you might ask them to explore one of the following questions:

- (Ask midway through a game) How many more counters will you need in order to get 15?

- Is there a way you can record what happens in the game, so that you could recount the whole game to someone afterward?

- Did you have more than 15 at the end of the game? How many more?

For additional challenges, see the extensions (p. 86).

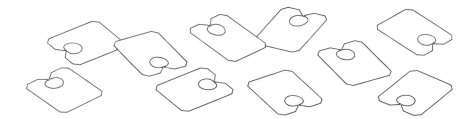

Session 1 Follow-Up

 Homework

Collect 15 Together Students take home Student Sheet 5, Collect 15 Together (the game directions), and teach someone at home how to play. They will need a dot cube (like the ones that come with many children's board games).

Students could make their own dot cubes, perhaps as a class project or at home, using small wooden cubes or half-pint milk cartons. To make a cube from a milk carton, first wash it thoroughly. Fold down the top so that it lies flat, and secure the folded top with tape. Cover the cube with construction paper or contact paper and draw the dots with crayon or marker.

Alternatively, students might use the Number Cards they took home in Investigation 2, and remove all cards except the numbers 1 through 6. They mix the deck, place it facedown, and turn over one card on each turn to determine how many counters to take. For counters, they might use buttons, beans, toothpicks, pennies, or paper clips.

Extensions

The game of Collect 15 Together has many possible variations to keep interest in the game high through many repetitions, and to add challenge as students become ready. You could also combine one or more of the following variations.

Collect Exactly 15 The goal is to collect exactly 15 counters. If a player cannot use the number rolled because there would be more than 15 counters, he or she may roll again (or, students may decide that the player loses a turn).

Collect and Record At the end of each turn, students write both the number they rolled and the total number of counters they have so far. For more challenge, students also record the number they need to get to 15.

Collect 25 Players play with the goal of collecting 25 or some other number of counters. If players are collecting a large number of counters, they could use two dot cubes or a deck of Number Cards 1–10 to determine how many counters to take on each turn.

Collect with Two Dot Cubes On each turn, players roll two dot cubes. They may use one or both of the numbers that are rolled. For example, if they are playing to collect exactly 15, they may sometimes want to use only one of the numbers.

Collect with Cards At the start of the game, players remove the wild cards from a deck of Number Cards, shuffle the deck, and turn it facedown. On each turn, a player turns over the top card in the deck, reads the number on the card, and collects that many counters.

Collect with Wild Cards Players put the wild cards back in the Number Card deck. When they draw a wild card, they assign it the number needed so their collection will contain exactly 15 (or 25) counters. This version can also be played with dot cubes; choose one face of a cube and tape over it with masking tape to create a "wild card."

Counting and Combining

Materials

- Number Cards
- Counters such as buttons, bread tabs
- Dot cubes (1 per pair)
- Pattern blocks: 1 bucket for each 6–8 students
- Geoblocks: class set
- Interlocking cubes: at least 30 per student
- Game Record Sheet (1 per student, homework, optional)

What Happens

During Choice Time, students work on three activities: Collect 15 Together (or a more advanced variation of the game); What Comes Next? with their choice of materials; and either of the card games learned earlier, Compare or Double Compare. Their work focuses on:

- counting a group of objects
- keeping track of the size of a growing collection of objects
- combining two quantities
- comparing two numbers to find which is larger
- predicting what comes next in a pattern sequence

Activity

Choice Time

Post a list of the three choices for Sessions 2 and 3. Briefly review each choice and explain that students will be working on these choices for most of the next two sessions. By the end of that time, they must have done each of the choices at least once.

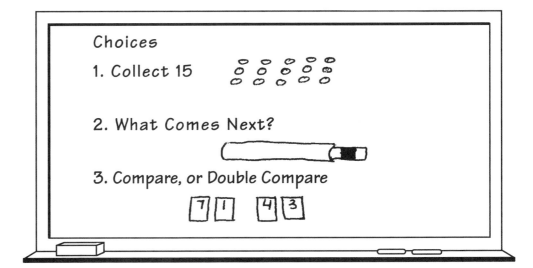

You may want to modify this choice list to reflect particular choices you are offering. For example, with Choice 1, you may also want to list a more advanced variation of the game that you have introduced (or plan to introduce). For Choice 3, you might list only Double Compare if you are no longer offering Compare.

Choice 1: Collect 15 Together

Materials: Dot cubes; a supply of counters

Students continue to play Collect 15 Together, as introduced in Session 1. If you noticed some students having difficulty working with 15 counters in Session 1, suggest they work with fewer counters, such as 10 (that is, they play Collect 10 Together). Likewise, if you introduced a more challenging variation of Collect 15 Together to some students in Session 1, they continue playing that variation; or, introduce yet another variation of the game to them (refer to the Session 1 extensions, p. 86).

Choice 2: What Comes Next?

Materials: Pattern blocks, interlocking cubes, Geoblocks, counters, sheets of paper or cardboard to cover the pattern sequence

This version of What Comes Next? is similar to the one that students did in Investigation 3 (p. 68), when they used pattern blocks. Here, they may choose from any of the available materials to create their pattern.

Each student, working at a different table from his or her partner, makes a pattern with about 12 things, laid out in a single row. A student might use all one material, such as Geoblocks, but could also use different materials to make a pattern. For example:

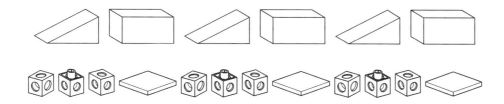

When the sequence is finished, the student covers it with a sheet of paper or cardboard or a book, leaving only the first two or three objects showing. Students take turns trying to predict their partner's sequence, one object at a time. As each new object is revealed by sliding the covering sheet back, the partner continues to guess what the next object is until the entire sequence is determined.

Choice 3: Compare or Double Compare

Materials: Decks of Number Cards, with wild cards removed (1 per pair); cubes or other counters (available)

Students play Compare in pairs, dealing out a deck of Number Cards evenly and turning over their top card simultaneously. The player whose card shows the larger number says "Me." If the numbers are the same, players mix these cards back in their pile and turn over the next card. The game is over when both players have turned over all their cards.

Double Compare is played the same way, except that students turn over two cards at a time and say "Me" if their cards make a larger total.

Observing the Students

As students work on the three choices, watch for the following:

Collect 15 Together

- How accurately do students count the counters? What strategies do they use for counting? for keeping track of what has been counted and what needs to be counted?

- How do they find the total number of counters after the new ones have been added?

- Do they play cooperatively and help one another with the game?

- Do they recognize when they have reached 15? Do they recognize when they have reached a number greater than 15?

What Comes Next?

- Are students making a variety of patterns (not just a-b-a-b)?

- When they are guessing their partner's pattern, do they analyze the materials showing to help them predict?

Compare or Double Compare

- Can students use the numerals on the cards, or are they counting the pictured objects? Do they count accurately?

- What strategies do they have for determining which number or total is greater? Do they "just know"? Do they count or count on? Do they reason about number combinations? ("I have more because mine are 6 and 3, and yours are 2 and 3, and 6 is more than 2.")

Sessions 2 and 3 Follow-Up

Math Games Students play one of this unit's games with someone at home. They might play Double Compare or one of the variations of Collect 15 Together. If you did not send home a copy of the Game Record Sheet earlier, you might include one with this homework. You could either specify one game or a particular variation at the top of the sheet, or give students their choice.

Homework

Ways to Make 15 After some experience with the game Collect 15 Together, students can further investigate the number 15, using a deck of Number Cards with wild cards removed. They may explore one or more of the following:

Extension

■ Find ways to make 15 with three cards. How many different ways can you find?

■ Use as few cards as possible to make 15. What is the least number of cards you can use?

■ Use as many cards as possible to make 15. What is the greatest number cards you can use?

For each question they investigate, they write the number of cards used and show how the cards make 15.

Students could explore these questions with a larger number, such as 25.

Eleven Fruits

Materials

- Counters (cubes, buttons, bread tabs) in at least two colors
- Student Sheet 6 (1 per student, homework)

What Happens

As an assessment, students solve a How Many of Each? problem about eleven fruits (or another number, as needed to adapt the level of difficulty to your class). Students then share their solutions and solution strategies in a class discussion. Their work focuses on:

- finding combinations of 11
- using pictures, numbers, and words to record solutions to a problem
- finding more than one solution to a problem

Activity

Assessment

11 Fruits: How Many of Each?

This assessment gives you an opportunity to observe how students find at least one combination of 11, or another number at an appropriate level of challenge for them.

You will need to decide what problem students solve. You have two variables to determine:

- *Total number of objects.* Choose a number, such as 11, that students have not yet worked with in the daily activities. You might ask all students to work with the same total, or you might decide that some students need to work with a smaller total, such as 6.
- *Which objects.* You might give students a general category, such as fruits, favorite foods, or pets, and let them choose what particular items they want to use to make 11. If this seems too difficult for them, you can assign two items, such as cherries and lemons.

If some students worked with three items when they did How Many of Each? problems earlier in the unit, you might also allow them to include some solutions with three items here. However, they should first find solutions for two items.

❖ **Tip for the Linguistically Diverse Classroom** If you choose the two items for students, students with limited proficiency in English will need a clear visual idea of what the items are, so that they can make their own representations as they record their solutions. Show pictures or examples of real items (for example, an apple and a banana) before doing this assessment.

Present the assessment problem orally; for example:

You have eleven fruits in your basket. Some are one kind of fruit, and the rest are another kind. How many of each could you have?

You need not write the complete problem on the board, but do record the total number as a reminder (for example, 11 fruits).

If you find that some students are most comfortable solving the problem with the familiar peas and carrots, do not insist they work with other items. After they have solved it, you can encourage them to think about how their answer compares to one that uses the same numbers but different items:

I see that you got 7 peas and 4 carrots. Andre solved the problem with strawberries and blueberries and got 7 strawberries and 4 blueberries. How is your solution like his? How is it different?

Students work individually. They may use counters to help solve the problem. When they have a solution, they record it on paper, using pictures, numbers, words, or a combination of these.

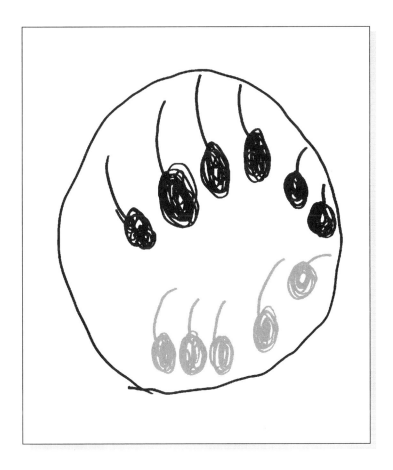

Observing the Students

As students are working, circulate to gather information about how each student is approaching the problem. Don't hesitate to ask them to tell you how they are going about solving the problem, to ask them to count or recount sets of objects, or to explain their recording methods to you. However, avoid telling them if their work is correct or incorrect. Look for the following:

- How do students approach the problem? Do they use counting? number combinations? Do they seem to work strategically, or do they use trial and error?

- How do students model the problem? Do they use manipulatives? pictures? numerals? Do they work mentally?

- Can students count accurately? Do they count from one each time, or do they count on from a number? Can they combine two sets and determine the total? Can they keep track of the total number of items and the number of each type?

- What strategies do students have for checking that their solutions are correct?

- How do students record their solutions on paper?

- Do students recognize that the problem has more than one solution? Do they look for multiple solutions? Do they find relationships among the solutions?

Sharing Solution Methods

When everyone has found at least one answer, gather students together to discuss how they solved the problem. Students bring their papers to help them explain what they did.

As a few students share, record their answers, but keep the focus of the discussion on how they solved the problem. See the **Dialogue Box**, Solution Strategies for 11 Fruits (p. 98), in which students in one class are describing their solution strategies. Notice how the teacher encourages them to talk more about their strategies.

As you record solutions, borrow a form of recording that you saw as you watched students working. Use a different form than you did for the previous How Many of Each? problem (Seven Peas and Carrots, p. 48). Review the **Teacher Note**, When the Teacher Records Students' Solutions (p. 53), for suggestions.

Each time someone shares a new strategy, ask if anyone else solved the problem in a similar way. Some students may say that their strategy is similar because their answer uses the same numbers (8 apples and 3 oranges; 8 strawberries and 3 blueberries) or the same two items (8 strawberries and 3 blueberries; 4 strawberries and 7 blueberries). In these instances, acknowledge that the answers are similar in certain ways (same numbers, or same fruits), and then encourage students to tell how they found their solutions.

See the **Teacher Note,** Encouraging Thinking and Reasoning (p. 97), for suggestions on helping students to explain their strategies and recognize what's important to relate when explaining how they solved a problem.

At the end of the session, collect all the papers so that you can learn more about how students record their work.

Session 4 Follow-Up

Homework

Dinosaurs and Tigers On Student Sheet 6, Dinosaurs and Tigers, students solve another How Many of Each? problem and record their solution or solutions. Depending on your class, you might assign the same problem to everyone, or you might ask different students to work with different totals. As needed, modify the number in the problem (for example, changing 10 dinosaurs and tigers to 8 or 15). Students ready for more challenge can try to find all possible solutions.

❖ **Tip for the Linguistically Diverse Classroom** Show pictures or make simple drawings of the items in the combination (dinosaurs, tigers).

Note: Students need not always work with a different total each time they do a How Many of Each? problem. Repeating the same total with different objects gives them more practice with number combinations.

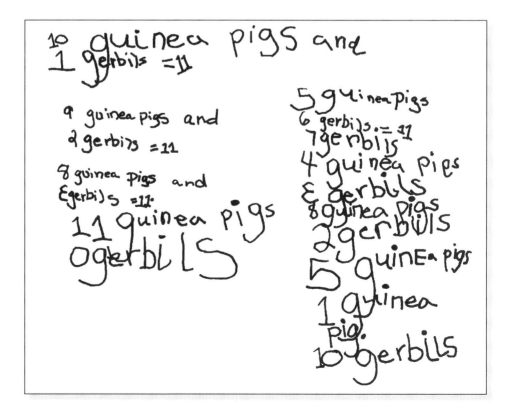

Encouraging Thinking and Reasoning

Students need to take an active role in mathematics class. They must do more than get correct answers; they must think critically about their ideas, give reasons for their answers, and communicate their ideas to others. Reflecting on one's thinking is a challenge for all learners, but even the youngest students can begin to work on this important aspect of mathematics.

Teachers can help students develop their thinking and reasoning by asking them "How did you find your answer?" or "How do you know?" If these questions evoke responses such as "I just knew it" or no response at all, you might reflect back something you observed as they were working. "I noticed that you made two towers of cubes when you were solving this problem." This gives students a concrete example they can use in thinking about and explaining how they found their solutions.

By asking questions or by reflecting on what a student has said in a whole-class discussion, you engage not only that student, but also other members of the class in thinking about an answer or statement. After one student has explained his or her solution method, asking "Who found the answer in a similar way?" extends the discussion and encourages students to think about similarities among strategies. Asking "Who found an answer a *different* way?" again extends the discussion and encourages other students to share their ideas. Because you will not usually have time for each student to share a solution method, this also serves as a way to acknowledge the approaches of many different students.

As time permits, ask students to explain why they think their approaches are like that of another student. This helps students clarify and articulate their ideas, and also helps them learn to distinguish between *a solution* and *a method for finding a solution*. Some students may say that they found a solution in the same way as another student because both arrived at the same answer: "Tony and I found the answer the same way. We both got 6 apples and 5 bananas." In these instances, acknowledge that the solutions are similar, and then encourage students to tell how they got them. Keep in mind that some students may be ready to attend to similarities among solution methods, but others may need to focus on articulating their own approaches. They may not be ready to compare their strategies to those of their classmates. What's most important in first grade is that all students recognize that there can be many ways of finding solutions to a problem.

The ability to reflect on one's own thinking and consider the ideas of others evolves over time, but even young students can begin to understand that an important part of doing mathematics is being able to explain your ideas and give reasons for your answers. Over the year, your students will become more comfortable thinking about their solution methods, explaining them to others, and listening to their classmates explain theirs.

Solution Strategies for 11 Fruits

For the assessment in Session 4, students in this class solved the problem "I have 11 of two kinds of fruit. How many of each do I have?" Each student decided individually which fruits to use. A few chose to use three (as permitted by the teacher), and a few used peas and carrots. Here are excerpts from their follow-up discussion.

Now we're going to share our work on the eleven fruits problem. Who wants to tell us how you solved the problem? Not just how many of each fruit you got, but also *what you did* to get your answer.

Mia: I did one melon and one lemon, one melon and one lemon and then I counted. I kept on doing it and counting them.

How many melons did you have?

Mia: *[counting the orange circles on her paper]* Six melons.

How many lemons?

Mia: *[counting the yellow circles on her paper]* Five lemons.

Did anyone do it like Mia?

Luis: I made one apple and one pear, over and over, until I got to 11.

Yes, that's similar to the way Mia did it. How did you know when you got to 11?

Luis: I don't know.

What did you draw first? Did you draw an apple?

Luis: An apple and a pear. Then another apple and a pear. *[He stops to count.]* I did five apple and pears. Then another apple.

How many of each did you have?

Luis: Six apples and five pears.

Chanthou: I did it like them! I had five peas and six carrots.

We have the numbers five and six again. What did you do first to get your answer?

Chanthou: I drew five peas.

So you started out differently than Luis and Mia, but you got the same two numbers they did. How did you know how many carrots to draw?

Chanthou: I counted in my head and I used a pencil so I could erase.

Chanthou started out by drawing some of one thing. Then she counted to see how many more she needed. Did anyone else do something like that?

Jacinta: I got peas and carrots, too.

The same things. What did you do first?

Jacinta: I drew *[she counts green circles on her paper]* three green.

And then what did you do?

Jacinta: Then I did *[she counts circles again]* three orange and *[counts]* four green and the rest orange, and I counted how many, and I put numbers under them, and I had seven green and five orange.

Nathan: But 7 and 5 is 12.

What do you think, Jacinta? *[Pause.]* Why don't you double-check?

Jacinta: Uh, I have... *[counting all the circles on her paper]* I have twelve.

Take a few minutes and try to figure it out. Let us know what you find. While you're thinking, let's hear from someone else.

Nathan: I know 8 and 3 is 11, because 9, 10, 11, so I put eight raspberries and three peas.

Nathan already knew that 8 and 3 is 11. Did anyone else use what they knew about two numbers that add together to 11?

Chris: I knew in my head that 10 and another number is, like, that number, like 10 and 1 is 11. So, I made ten blueberries in a circle and one strawberry in the middle.

Yukiko: I did circles like that. I did a banana with blueberries around it, and then I counted and I didn't have enough. So on the side I drew another banana with raspberries around it, and then I had too many, so I crossed some out until I had 11.

How many of each did you end up with?

Yukiko: Two bananas, five blueberries, and four raspberries.

Jacinta *[calling out]:* I figured it out now. I crossed out one of the peas 'cause I had 12. There's still five carrots and there's *[she counts]* six peas.

Good job on sticking to the problem, Jacinta.

Making Predictions About a Story

Materials

- *Rooster's Off to See the World* or a similar story

What Happens

Students listen to the first half of a story in which a total grows by accumulating progressively larger groups (such as *Rooster's Off to See the World*). They use pictures, numbers, and words to show what they think happens next in the story. A few share their ideas with the class. Their work focuses on:

- counting a set of objects
- exploring the concept of "one more" than
- making and explaining predictions
- exploring a numerical pattern

Activity

A Story with a Growing Pattern

The picture book *Rooster's Off to See the World* illustrates a growing pattern that increases by one number at a time. In the first half of the book, different groups of animals join Rooster on a journey to see the world, each group being one larger than the previous group: first two cats join, then three frogs join the rest, then four turtles, and finally five fish. In the second half of the book, each successive group tires of the journey and goes home, in reverse order (first the five fish, then the four turtles, and so on), until only Rooster remains.

If you have this book, introduce it and read aloud the first half. As you are reading, you might pause once or twice to ask students to predict what they think happens next. For example, after reading that three frogs have joined the group, ask students what they think happens next. Accept two or three responses from students, and then continue reading. Stop reading after the five fish join the group, and go on to the next activity.

If You Can't Get This Book Many children's books illustrate the pattern of steady accumulation of "groups of one more." Any such story will work (it need not go up to five and then back down, like the Rooster story).

One good choice would be *P. Bear's New Year's Party* by Paul Owen Lewis. In this story, P. Bear invites all his best-dressed friends to a party. Successively larger groups of animals arrive each hour, from one to midnight.

Another alternative is *The Very Hungry Caterpillar*, also by Eric Carle; on successive days, the caterpillar eats larger and larger numbers of things. Because children may already be familiar with this book, the "predicting" part of these sessions may become more of a "tell what happens next" activity. Even if students know the ending, this is still a good book to use for Session 6, How Many in All?

Whichever book you use, read partway through it in this session, so that students have a chance to hear about several increasingly larger groups. For example, if you are reading *P. Bear's New Year's Party*, you might read up through 5:00, when 1 whale, 2 horses, 3 cows, 4 zebras, and 5 pandas have arrived at the party. Then, stop and ask students to predict what happens next.

If you cannot get a suitable book that illustrates this pattern, you can easily make up your own story. Use characters or objects familiar to your class for added interest. Begin with one animal or thing, and invent a situation in which you add a group of two, then a group of three, then four, then five. For the second half of the story, you can either continue the pattern upward or reverse it and return to one (as in Rooster's story); what's important for these sessions is having successively larger groups from one to five.

What Could Happen Next?

Explain to the class that you will be reading the rest of the story to them tomorrow, but that today you would like them to predict what might happen next. If some students have heard this story before, ask them to keep the ending secret.

To help students think about what could happen next, reread the first half of the book to them, asking them to listen carefully. Do not point out the pattern or make a list of the numbers or types of animals that join Rooster, as you will want to observe what students include in their predictions and remember on their own.

When you have finished reading, ask students to write or draw something to show what could happen next. Students who know the actual ending to the story can show it, or they can make up a new ending to the story.

Observing the Students

While students are working, circulate to look for the following:

■ Do students include any numeric information in their stories? Have they remembered the numeric information from the book?

■ How do they represent any numeric information? with pictures? words? numbers?

■ Do they appear to recognize a pattern in the numbers of animals that join Rooster? Do they try to extend the pattern in their stories?

Note some of the different predictions and who made them; you might focus on some of these during the follow-up discussion.

Expect wide variation in students' ideas about what happens next. Some students may describe the journey that the animals take ("They built a boat and began sailing around the world"); others may recognize and extend a pattern ("Next they met six goats"). Still others may include numeric information that does *not* extend the pattern ("Then they met two dogs"); perhaps they did not recognize the pattern, or perhaps they did, but chose not to extend it.

"Two dogs will walk with them."

This activity provides a context in which students *might* recognize and represent numeric information or a numeric pattern. However, a student response that includes numeric information should not be considered preferable to one that includes no numeric information. The directions are intentionally open-ended, and different students will focus on different aspects of the story. This predicting activity gives first graders the time they need to think about the story and the characters before they start problem solving with the information. Wait until the next session to explicitly turn students' attention to the numeric information in the story.

Students who finish early can quietly explain to a partner who has also finished what they think happens next.

"They climbed a ladder and they got stuck on the moon and the fish are lucky and they went home."

Sharing Predictions About 15–20 minutes before the end of the session, call students together, with their papers, to share some of their ideas about what could happen next.

Remind students that anyone familiar with the book should keep the ending secret until tomorrow, when you will be reading the whole book to the class. Explain that you would like to hear from students who are not familiar with the book.

Ask for volunteers to hold up their papers and explain their ideas to the class. After each student shares, ask for others to compare their own predictions to that one.

Fernando predicted that the animals would all take a boat ride. Did anyone else predict something similar? Did anyone predict anything different?

To ensure variety, you might call on students who (based on your observations) have made different predictions. For example, you might call on someone who predicts that Rooster and his friends will next meet a group of six animals; you might call on a student who predicts that they will discontinue their trip.

See the **Dialogue Box**, Predicting What Happens Next (p. 105), for excerpts from one class discussion.

"We saw 6 squirrels. 'Would you like to come with us?' And the squirrels went."

Predicting What Happens Next

These students are sharing their ideas about what could happen after the five fish join Rooster in *Rooster's Off to See the World.* Throughout, the teacher encourages students to explain how they arrived at their ideas, since making predictions *and* explaining one's reasoning are both very important mathematical activities.

What do you think could happen next?

Claire: Seven octopuses.

What are they going to do?

Claire: Play. Jumping in the water. Grab people.

What makes you think seven?

Claire: Because that's how many there were. One octopus and six friends.

Other sevens? No? OK, let's hear another idea.

Nathan: Six of these things *[his paper shows six creatures in water]*. They're kind of a seal kind of thing. They swim. They're the same shape as a seal. They're white and they have dots on them.

What happens to your seal things?

Nathan: The rooster's going to ask them please come. They'll say yes. Might see the world.

Why six?

Nathan: Because the fish were five. Because we went to five and now it's time for six to come.

Anyone else have six of something? *[Several hands go up.]*

Tony: I got six gold dogs. They're going to travel around with Rooster.

How do you think they'll travel?

Tony: They're gonna walk.

Why six?

Tony: Because they want to go with the rooster and they said please. They want to play with each other.

But what made you think six dogs? *[Pause.]* **Not sure? OK, if you think of why you did six, just raise your hand and let us know. Who else?**

Nadia: Six alligators.

What are these boxes you drew?

Nadia: The boxes from the book. Like, one rooster. Two cats. Three frogs. Then four turtles. Five fish. Then six, now I think six alligators.

Why do you think there will be six?

Nadia: They're numbers. 1, 2, 3, 4, 5, 6.

We've had seven octopuses, and six of a few different animals. Who had another idea?

Libby: All the animals will go running back to the barn together.

Why do you think they might do that?

Libby: Because it is dark and they're scared.

Sometimes it helps when you're scared to be with someone. Did anyone else have an idea like Libby's? Something different?

Tamika: Five dogs.

Why five dogs?

Tamika: Because we had five fish. Need something new.

Other fives? Something else?

Yanni: Ten caterpillars. They're gonna wait on the porch for the rooster.

What made you think ten?

Yanni: Because I saw their tails when the page [of the book] was flipping.

Other tens? Who has another idea?

Brady: Two roller skates.

What's going to happen?

Brady: Everyone is going off on roller skates with the rooster and then the sun came up.

Session 6

How Many in All?

What Happens

Students listen to the rest of the story that was started in Session 5 and determine how many in all are in the story. They record and share their work. Their work focuses on:

- using a variety of methods, such as pictures, concrete materials, and numbers, to represent information about a series of groups
- combining several quantities
- using counting, patterns, and other strategies to help solve problems

Materials

- Book or story from Session 5
- Cubes or counters

Activity

Teacher Checkpoint

How Many in All?

This checkpoint gives you an opportunity to observe how students represent information about a series of groups, and what strategies they use for combining several quantities.

Note: If you are using *Rooster's Off to See the World*, students will be combining the quantities one through five. If the book you are reading or story you are telling involves groups of more than five, pose a problem based on the stopping point of five groups. Students might later determine, as an extension, the total number after a group of six or more has joined.

Yesterday we read the first part of *Rooster's Off to See the World*, about Rooster and his friends. What happened in the story so far?

Ask a volunteer to summarize the story so far. Other students may remember the story differently; allow brief discussion of any disagreements.

We stopped reading when five fish joined Rooster, and you made a lot of interesting predictions about what could happen next. Now, we're going to find out how the story ends.

Finish the story you have been reading. Then invite a few volunteers to share their thinking about the ending of the book. Were they surprised? Do they like the ending? Does the ending match their predictions?

Introducing the Problem Explain the task that students will be working on for most of this session.

When Rooster went on his journey, a lot of groups of animals joined him. Today we'll be figuring out how many animals there were altogether.

Before students begin working on this task, you might reread the story up through the point at which you have five groups. You might also ask for volunteers to tell you the kinds of things in each group and the order in which they appeared. For example:

What kind of animals first joined Rooster? Who joined after the cats?

On the board or a piece of chart paper, draw a picture to show the types of things in each group, in order. Thus, for Rooster's story, you would draw a cat, a frog, a turtle, and a fish. Do not list the *number* in each group, as you will want to see if students are able to recall this themselves, or if they notice a pattern.

You might let students decide for themselves whether or not to include Rooster in the total count of animals. Some students may decide to do both: find the total with Rooster and without Rooster. Students may work alone, in pairs, or in small groups. They may use counters or cubes to help them. Encourage students to share their ideas with one another. When they have found a solution, they record it using pictures, numbers, words, or a combination of these.

Put the book away as students are working on this problem; they should not be able to look at the pages that show the growing groups, because it is important for them to develop their own ways of representing the groups and finding the total number.

Observing the Students

Circulate to observe how students approach the problem and to offer support as needed.

■ Do students remember how many animals in each group?

   If some students are having trouble remembering the numbers, ask them to tell you again what happened in the story. How many animals did Rooster meet first? then how many? how many after that? Some students may begin to recognize the pattern. If some still cannot remember the number in each group, ask a classmate working nearby to remind them, or simply tell them yourself.

■ Do students have a way to represent the number in each group? Do they use concrete materials? pictures? numbers? a combination?

- Do they arrange their representations in an orderly way (for example, from the smallest to largest group of animals, or with each type of animal in a different row)? Do they appear to notice a pattern?

  To help students who are having difficulty, ask questions like these: Suppose you wanted to tell someone else the story. What could you do to help you remember the animals that joined Rooster? How could you keep track of the *type* of animal in each group? the *number* of animals in each group? Could you use counters or draw something to help you? If some students record only the total number of animals, ask them to show how many in each group. For example, if a student has drawn a tally mark for each animal but not arranged them in groups, say: "Show me on your paper which tally marks stand for which animals." If a student has simply recorded a number, ask for a picture showing the number of animals in each group.

- How do students go about finding the total number of animals? For example, do they count the number of pictures in their representation? Do they find the total of the numbers they have recorded? Do they use patterns to help them find the solution? Do they use counters or their fingers to help them find the total?

- What strategies do students have for checking that their solutions are correct?

- How do students record their answers? Do they use pictures, words, numbers, or a combination? Do they use numbers or number words to show "how many"?

When students have finished, they explain their solutions to a partner. If some partners get different answers, encourage them to think about why their answers differ. Did they include the same groups of animals? Did they count groups of the same size? Do they need to double-check their work?

Note: Students who are comfortable with calculators may want to use this tool to find the sum of the numbers 1 to 5. This is fine, but encourage these students to think about whether answers they obtain on the calculator are reasonable. Also insist that any students using the calculator record how they are solving the problem, just as they would if they were using cubes, mental addition, or some other method to arrive at their answers. If some students record only the total number, ask them to tell you how they arrived at their answer, and then to show on paper what they did. Some students may record the numbers they added on the calculator; others may draw a picture showing the number of animals in each group. Students who use the calculator should also try to find the total in a different way.

Students who finish early could try one of the extensions (p. 110). For example, in Rooster's story, how many animals are still there after the fish leave? after the turtles leave?

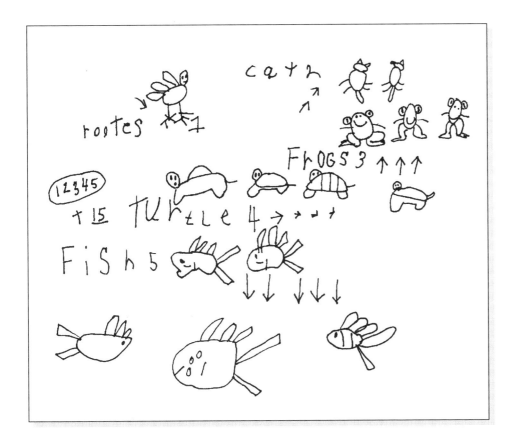

Sharing Solutions When everyone has found a solution, gather students together, bringing their papers, to share some of their solutions.

How did you get the total number of animals? Who has a way to share with us?

Students who share should tell (1) how they kept track of the groups (with cubes? pictures? numbers?), and (2) how they found the total. If some students in the class have counted Rooster in the total and some have not, be sure that those sharing tell whether Rooster was counted or not.

Kristi Ann used cubes of a different color for each kind of animal, then she counted up all the cubes. Did anyone else find a solution that way? Did anyone find the solution a different way?

Call on several students to share their ideas. To demonstrate a variety of approaches, look for volunteers who organized their work and solved the problem in different ways. For example, if one student drew and counted pictures of each successive group (from one rooster up to five fish) to find the solution, you might call on a student who also used numbers.

If any students used equations, encourage them to explain their work, but treat equations as just one of the many good ways of recording that students have used. For young students, finding a way to record their thinking and their solutions is an important part of making sense of problems. At this point in the year, it is critical not to emphasize one particular recording method over the others. See the **Dialogue Box,** How Many in All? (p. 111), for examples of students' recording methods, including equations, and ways that students explained their work.

Session 6 Follow-Up

Extension

Exploring Other Totals in the Story Students ready for more challenge can explore one of the following questions for Rooster's story, or similar questions if you have used a different book:

■ How many animals in all were there at different points early in the story: after the frogs joined? after the turtles joined?

■ How many animals in all were there at different points later in the story: after the fish left? after the turtles left?

■ Imagine that another group of six animals joined Rooster after the fish joined. How many animals would there be in all? What if a group of another size joined?

Some students may want to use calculators to investigate these questions.

How Many in All?

These students heard the story *Rooster's Off to See the World*, and then worked to find out how many animals in all set off on the trip together. Here they are sharing their work in a follow-up discussion.

Let's talk about how you solved this problem. Listen closely to your classmates. Think about what they tell you. How did they find the answer? See if you did it the same way they did, or a different way. When you share, hold up your paper so everyone can see.

Luis: I did it with Jamaar. We used cubes.

Can you tell us more? Or show us what you did?

Luis: *[He arranges cubes in a staircase.]* One, then two, then three, then four, then five. Because one rooster, two cats, three frogs, four turtles, five fish.

Jamaar: I drew what we did a different way.

So how did you figure out how many went altogether?

Luis: Um ... Jamaar, I need your help.

Jamaar: I counted until I was at 15. We counted the cubes.

So they showed all the groups of animals with cubes, and then counted them up. Look and see if what you did was like what Luis and Jamaar did. Did anyone do something similar?

[Several students raise their hands.]

Let's hear from one person who did something similar. How was your way like their way?

Continued on next page

Yukiko: It was like theirs but I didn't use cubes. I got 15 too because I counted, all the way to 15.

What did you use to help you keep track?

Yukiko: I drew a picture for each animal.

Eva: I did that too, but I did it like in the book.

Did anyone else draw pictures of animals, and then count them all? *[Many hands.]* **Lots of you did! Let's hear from someone else.**

Iris: I drew pictures too. Then I drew lines and counted them.

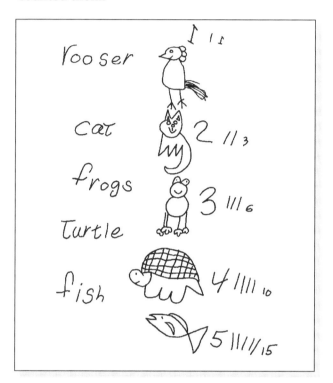

Some of you drew pictures to show the groups of animals, some of you showed them with cubes, some of you showed them with lines. Then, you counted them up. Did anyone do something different?

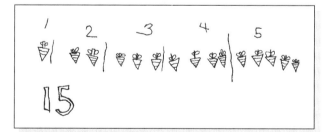

Michelle: I used carrots. Because from last night I did 15 in all for homework.

Can you tell us more?

Michelle: I did ten carrots and five carrots.

How did you know to draw 10 carrots and 5 carrots?

Michelle: Because if you added those numbers *[she indicates the 1, 2, and 3 on her paper]* it'd be six, and four more is ten. And then five more, and it's like ten carrots and five carrots. I did five and ten peas and carrots for homework.

The carrots and peas helped her think about this. Did anyone else think about carrots and peas? *[Pause.]* **Did anyone else do something similar to Michelle?**

Diego: I did numbers, too *[shows his work]*.

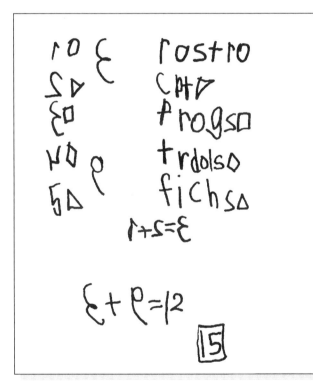

Can you show us how you put those numbers together?

Diego: *[Pointing to the 1 and 2 at the top left of his page]* One plus two is three. *[Next he points to the 4 and 5 at the middle left.]* Four and five is nine. *[Finally, he indicates the 3 + 9 at the bottom of the page.]* Then I put those numbers together and got... 20? Three and nine is... 20? [Diego speaks English as a second language and can't readily recall the name of the number.]*

Remember the name of that number? Anyone want to help?

Diego: Um, twelve?

Good for you. I know sometimes it's hard to remember the names of all these numbers. So, Diego found out that the 1 and the 2 together made 3, and he wrote that down here. Then he took this 4 and 5, put them together, and made 9. Then he put the 9 and the 3 together, and he wrote that at the bottom: 3 + 9 = 12. Then, Diego, how did you get to 15? Where did you get that?

Diego: I had to add in the 3, so I just counted, 13, 14, 15.

What a lot of ways all of you came up with to show how you thought about this problem! They're all good ways to show how many in each group, and good ways to help you find how many in all. You'll have a lot more chances to keep finding ways to show your work this year.

INVESTIGATION 5

Data About Our Class

What Happens

Session 1: Kid Pins Students find how many are present in their math class by counting around the class. They work individually to make Kid Pins, decorated clothespins that they will use throughout this investigation and later in the year to keep track of votes in classroom surveys.

Session 2: Attendance and Other Surveys Students do several quick surveys and use their Kid Pins to represent the data. In an attendance survey, they use Kid Pins to show how many are present and absent today, then find ways to use this data to determine the total number of students in the class.

Sessions 3 and 4: Inventing Survey Representations Working with another survey question, students are introduced to the idea of making representations of collected data. They first sort people, then use cubes or Kid Pins to represent the people, and finally make representations on paper with markers or stick-on materials.

Sessions 5 and 6: How We Got to School Today As students consider how they (as well as children around the world) get to school, they work with data that involve more than two categories. They brainstorm ways of representing their class data and work individually or in pairs to make their representations.

Routines Refer to the section About Classroom Routines (pp. 145–152) for suggestions on integrating into the school day regular practice of mathematical skills in counting, exploring data, and understanding time and changes.

Mathematical Emphasis

- Collecting survey data
- Inventing representations that show clearly what a survey is about
- Categorizing data and communicating those categories in a representation of the data
- Representing the sizes of different groups with pictures, numbers, or words
- Counting, combining, and comparing the sizes of different groups up to about 30
- Making sense of survey results and presenting them to others

What to Plan Ahead of Time

Materials

- Two sheets of heavy cardboard (perhaps from a box) or two sheets of foam core, each about 12 by 24 inches (for survey boards used throughout the investigation; see Other Preparation)

- Wooden tongue depressors: 1 per student (Session 1)

- Spring-clip wooden clothespins: 1 per student (Session 1)

- Fast-drying glue (Session 1)

- Chart paper or newsprint: 15–20 sheets (for teacher use throughout the investigation)

- Stick-on notes: a supply of various sizes, for labels and data representations (all sessions)

- Fine-tipped markers and other drawing materials (all sessions)

- Interlocking cubes: about 20 per student (all sessions)

- Large paper (11 by 17 inches): 3 sheets per student (Sessions 3, 5–6)

- *This Is the Way We Go to School: A Book About Children Around the World* by Edith Baer (Scholastic, 1990); also available in Spanish (Sessions 5–6, optional)

- Buttons, construction paper, or colored dot stickers for making representations (Sessions 5–6, optional)

Other Preparation

- Make two survey boards by preparing foam core or cardboard in two different colors. The boards should be strong enough to lean against a wall and big enough so the entire class can clip their Kid Pins to the edges. (Session 1)

- Make a Kid Pin by gluing a clothespin to a tongue depressor with the opening at one end (see picture, p. 117). (Session 1)

- Arrange the room so that students have space to move around and sort themselves into two groups. (Session 1)

- Prepare stacks or containers of cubes in two colors, with enough of each color for one cube per student. (Sessions 3–4)

- Prepare a class list as a one-column list of names. Duplicate 2 per student for use in taking surveys in Sessions 3–4 and 5–6.

- Duplicate the following student sheets, located at the end of this unit. If you have Student Activity Booklets, copy only those items marked with an asterisk.

  For Session 2

  Game Record Sheet* (p. 178): 1 per student (homework, optional)

  For Sessions 3 and 4

  Student Sheet 7, Oranges and Cherries (p. 171), or Student Sheet 8, Oranges, Cherries, and Grapes (p. 172): 1 of either sheet per student (homework)

  For Sessions 5 and 6

  Student Sheet 9, How We Get to School (p. 173): 1 per student

Session 1

Kid Pins

Materials

- Prepared Kid Pin sample
- Wooden tongue depressors (1 per student)
- Spring-clip wooden clothespins (1 per student)
- Fine-tipped markers
- Fast-drying glue
- Two prepared survey boards
- Interlocking cubes (available)

What Happens

Students find how many are present in their math class by counting around the class. They work individually to make Kid Pins, decorated clothespins that they will use throughout this investigation to keep track of votes in classroom surveys. Their work focuses on:

- counting up to the number of students in the class
- matching sets of the same size

Activity

How Many Are Here Today?

Gather students in a circle. Ask for a few volunteers to share what they know about how many students are in their class. Students may have a variety of ideas; for example, they may tell you the total number in the class, the number who came to school today, or the number in the room right now (which will be different if not everyone who came to school today is in the classroom for math). Keep interlocking cubes available for students who want to use these to count.

We usually have [28] students in our class in all. We will check how many are here today for math class by counting. The first person will say "one," the next person will say "two," and we'll keep going until everyone has said a number.

Let's all stand up. When I call on you, say the next number and then sit down.

Call on or point to students in a particular order, such as clockwise around the circle. If you think particular students may have difficulty saying the larger numbers in the count, you might call on them first, so that they supply the earlier numbers.

We counted up to [26], so there are [26] in math class today. If we count again right now, what number do you think we will end up on? Why?

Many students will probably recognize that the total for today will always be 26, no matter how many times you count. A few may not be certain. Explain that the class will count again to check.

Stand up again. This time, we will all count aloud together as I point at each one of you.

For this check, point around the class in a different order (perhaps counter-clockwise around the circle).

Activity

Making Kid Pins

Explain that for the next week or so, the class will be doing different activities that involve counting the students in the class. They will be doing surveys, or collecting information about themselves. Invite students to share any experiences they have had with *surveys*, either this year or in kindergarten. Summarize their ideas or remind them of surveys you may have taken. For example:

We take a survey when we collect information, or *data*, about people. We were doing a survey when we talked about how many wanted apple juice and how many wanted orange juice for snack. Remember, you voted.

For the rest of this year, we're going to use Kid Pins to keep track of how we vote in our surveys. *[Show the sample Kid Pin you made.]* **Sometimes we raise our hands to vote, but with Kid Pins, we can keep a record of the number of votes. When you vote, you'll clip your pin to one of these survey boards** *[demonstrate].* **Afterward, we will be able to tell who voted for what, since your name will be on your Kid Pin.**

Today we're going to make Kid Pins for everyone. How many Kid Pins do we need to make?

Students may realize that this number is the same as the number of students in the class, and others may count. If some are absent today, remind the class that those students will need Kid Pins, too.

Each Kid Pin is made from one clothespin and one flat stick *[show these].* **How many clothespins do we need for the whole class? How many sticks?**

Give students a little time to think about this. Most or all hands should be up by the time you call on students. Again, many students will see the relationship between the number of students in the class and the number of clothespins. To a few, it may not be immediately obvious.

Decorate your stick and your clothespin with markers. You'll need to write your name or initials on the stick. Only write on one side of the stick, because you'll be gluing on the other side. When you've finished decorating, you'll glue the clothespin on like this.

Hold up your sample and draw attention to the parts where the glue will be placed and how the two pieces fit together.

Students take the rest of the session to complete their Kid Pins.

About 5 to 10 minutes before the end of class, remind students to begin cleaning up. When they are finished, they clip their pins randomly to the two survey boards, for storage. If some students finish early, they might figure out how many Kid Pins have already been made, and how many are still being worked on.

Every student needs a Kid Pin, so find a time for any absent students to make their Kid Pins when they return to class. You might use your own sample Kid Pin to participate in some or all of the upcoming surveys.

Attendance and Other Surveys

What Happens

Students do several quick surveys and use their Kid Pins to represent the data. In an attendance survey, they use Kid Pins to show how many are present and absent today, then find ways to use this data to determine the total number of students in the class. Student work focuses on:

- counting the number of students in different categories
- combining the number of students in different categories to find the total
- representing data that fall into two categories

Note: If no one is absent today, you'll need to do the attendance survey on another day. If this is the case, the session will be shorter than usual.

Materials

- Survey boards
- Kid Pins
- Interlocking cubes (available)
- Chart paper
- Game Record Sheet (1 per student, homework, optional)

Activity

Tell students that today they will get to use their Kid Pins for some quick surveys of the class. For this activity, you might use the following survey questions:

- Are you wearing a zipper or not?
- Are you a boy or a girl?
- Are you wearing sneakers or another kind of shoes?
- Do you have stripes on your socks or not?
- Do you like *pepperoni* pizza or not?

You can also devise your own questions, but be sure that they involve two clear categories. For example, wearing a zipper or not wearing a zipper involves two categories, but a survey about hair color involves more than two colors. See the **Teacher Note**, Data Categories (p. 125), for more information about considering data questions.

Quick Surveys

❖ **Tip for the Linguistically Diverse Classroom** Be sure to show an example or a picture of the two choices in the survey, so all students understand what is being asked.

Our first survey will be about who is wearing a zipper today and who isn't *[or another question of your choice].*

Ask students to come up in small groups and take their Kid Pins off the survey boards. Remove those belonging to any students absent today.

If you *do* have a zipper in the clothes that you are wearing, put your pin on the [white] board. If you are *not* wearing a zipper, put your pin on the [blue] board.

Students come up a few at a time and clip their Kid Pins on the board that represents the appropriate category. You may want to label each board with a large stick-on note, using pictures and words, to help students keep track of which is which. As students finish, ask them to look at their results.

Which looks like it's more—the number who don't have zippers today, or the number who do? How can you tell?

Students may be able to tell at a glance which category has more, especially if there is an uneven distribution of data. They may be looking at groups of pins, or at how much space is left, or at how much of each board is covered. While this is not a fully reliable analysis, it is a good way to start looking at the data. See the **Dialogue Box**, Making Sense of Kid Pin Data (p. 126), for a discussion that occurred in one class.

How could we check to see which is more: the people wearing zippers or the people not wearing them?

Students may suggest counting the Kid Pins, or counting the actual students. If you decide to count the Kid Pins, ask them first to predict how many Kid Pins will be in each group, then count. If everyone can see the survey boards easily, you might ask them to count on their own and then ask a few volunteers to share their results. If you think some students will have difficulty counting the number of pins without touching or coming very close to them, you (or a student) might lead the class in counting together, touching each pin as you count it.

Record the survey results on a piece of chart paper.

SURVEY RESULTS

14 zippers   11 no zippers

What did we learn from this data? What can you tell about the size of the groups? Were a lot more kids wearing zippers? Was it pretty close?

Students may observe (in our example) that more students have zippers. They may even be able to describe qualitatively that a lot more kids (or a few more kids) have zippers than don't have zippers.

How many Kid Pins are there in all? How do you know?

Encourage a few students to explain their strategies for finding the total. For example, they might count each pin (by counting all or counting on); they might remember the total number of students in class today from the earlier counting activity; or they might combine the two numbers.

Jonah said that there must be 25 in all because that's the number in math class today. Did anyone else think of it like that? Who did something different to find the total?

Once several students have shared their strategies and the class has agreed on the total, record it next to the results of the zipper survey.

SURVEY RESULTS

14 zippers   11 no zippers   25 total

Take at least two more quick surveys in the same way. Use questions that you think would interest your students; just be sure they have only two clear answers. There's no need to discuss each question as thoroughly as you did the first one, but do record the results of each survey on your chart, below the previous survey results:

SURVEY RESULTS

14 zippers	11 no zippers	25 total
10 boys	15 girls	25 total
18 sneakers	7 not sneakers	25 total

Then direct students' attention to the collected survey results.

We found that we always had a total of 25. Why do you think that is?

Some students may say that the total is always the same because there are that many students in class today. Others may recognize that the numbers in the first two columns show different ways of making up the total number. If some students seem ready, you might ask them if they can think of two other numbers (not already on the chart paper) that add together to make the same total.

Activity

Attendance Survey

Note: If no one is absent today, save this activity for the next day that at least one student is absent and simply end this session early.

In the surveys we've been doing, you've been clipping your Kid Pins to both of our survey boards. This time, I'd like all of you to clip your Kid Pins to the [blue] board. This will show us how many students are in class today.

When students have finished, ask for a few predictions about how many pins are clipped to the boards. As a group, count the Kid Pins.

There are 25 Kid Pins up here because there are 25 children in class today. We're going to use the [white] board to show how many are absent today. Who is absent today?

As children name absent classmates, clip each of these Kid Pins to the other board. (You can use a plain clothespin for any absent students who have not yet made their Kid Pins.)

When all the absentees have been accounted for, ask how many children are absent; then count the "absent" pins with the class. Record both numbers on chart paper:

```
                          ATTENDANCE

      October 9          25 here          3 absent
```

We found that 25 children are in class today, and 3 children are absent. How can we use these data to figure out how many children are in our class altogether?

Students may already know how many are in their class, but explain that you'd like them to use the survey results to find out in a different way. They may work alone or with a partner to find a solution.

Who would like to share their strategy?

Some students might count each pin or use interlocking cubes to represent each pin. Others might count on from the number of students present, and a few may suggest combining the numbers in some way. ("I knew 3 + 5 is 8, so 25 + 3 is 28.") Try out a few strategies that students suggest. For example, the class might count all the Kid Pins while the student who suggested this points to each pin. Or, if someone suggests counting on, ask that student to demonstrate for the class.

What do you notice about the total?

If someone notices that the total number is still the same, no matter what way you count, ask why this is. When the class has agreed on the total, record it on the chart paper.

```
                          ATTENDANCE

   October 9        25 here        3 absent        28 total
```

What if tomorrow, all three kids who were absent today are absent again, plus Jacinta is absent, too? Then what would the Kid Pins look like? What would we write on our chart paper?

Students may describe Jacinta's Kid Pin moving over to the "absent" board. Or, they may look at the numbers and calculate "one more" than the previous number of students absent.

On a new sheet of chart paper (headed What If?), record the new number sentence.

WHAT IF?

October 10   24 here   4 absent   28 total

Do a few more "what if" questions. For example, what if the same three, and Jacinta, and now Max are all absent? See what trends students observe.

What about the total now? Is it the same, or does it change?

Students share their ideas about whether and why the total number of students stays constant. Add the total to your chart paper record.

Post the first chart (with today's attendance data) in a convenient place, as you will be adding to it during later sessions.

Session 2 Follow-Up

 **Homework**

Math Games Even though your class work has turned to the exploration of data, students will benefit from continuing to play at home the games they learned earlier in the unit. Double Compare offers good practice with number combinations, as does Collect 15 Together. You might suggest one of the variations listed on Student Sheet 5, Collect 15 Together.

Consider sending home a Game Record Sheet, asking students to report on what game they played, with whom they played, and what happened. They can get help from an adult or the other player in writing this short report.

 **Extension**

More Quick Surveys Do some other quick surveys another time during the same day or on a day when the total number of students present is different.

Data Categories

During this introductory data investigation (Sessions 1–4 of Investigation 5), students work primarily with two categories of data. This is a good way to begin looking at survey results, since comparing two groups allows first graders to focus on the size of the groups. (In Sessions 5–6, they will use three to five categories to describe the ways they got to school.)

You may want to devise your own data questions that you think will especially interest your class. For the Quick Surveys (p. 119), consider carefully the possible responses. You will need a question that has exactly two answer choices. Consider patterns like these:

- Do you prefer _____ or _____?

- Are you _____ or NOT _____?

- Do you like _____ [bananas]? (Yes or no.)

Consider questions of fact about something that happened that day at school; for example, Did you jump rope at recess or not? Did you bring a backpack today or not?

Doing a survey about hair color, pets, or favorite ice cream flavor may be interesting, but is sure to involve many categories of data. Even a question that seems straightforward may be complicated. Think about this one: Are you wearing long or short sleeves? It seems simple, but what about students in sleeveless shirts? What about long *and* short sleeves (one shirt over another)? What about three-quarter-length sleeves?

Also consider whether your data question might raise sensitive issues. In one class, students talked about whether or not they'd had a bath or shower the previous night. Almost all students said they had bathed, probably because of peer pressure. Questions about abilities (Can you swim or not? Can you ride a two-wheeler or not?) have similar problems. Try to avoid questions about possessions (Do you have a _____?), which could be difficult for children who do not have these things.

Ideally, too, data questions will yield a good mix of both answers. Whenever there is a preponderance of one answer, young students are often tempted to "go with the majority" rather than respond honestly; when this happens, the data is less interesting.

If you would like to invite individual students to come up with their own data question, keep in mind that at this point, such a task is very difficult. Students will be asked to do this later in the process of learning about data. They may, however, be comfortable brainstorming topics with you in a group, if you want to include them in this aspect of data analysis.

Making Sense of Kid Pin Data

This class is doing Quick Surveys (p. 119). They describe the results of their first survey by comparing the sizes of the groups, then figure out the total number of students who participated. (They refer to their Kid Pins as *clips*.)

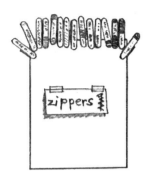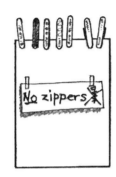

Let's look at what we found out. Those of you who have a zipper on your clothes today put your clips on the red board here. Those who don't have a zipper put their clips on the blue board. Which has more clips?

Leah: Zippers. Because there are more clips there.

What tells you?

Leah: I counted.

Anyone have another way of knowing which has more clips?

Susanna: *Zippers,* because they are more smushed together. The board's filled up. It's not like that one ["No zippers"]. That has more room.

Susanna knows because there is more room in between the clips here ["No zippers"] and hardly any empty space on this "Zippers" board.

[Together the class counts the seven clips for kids with no zippers.] **How many should there be on the other board?**

Max: Fifteen. I counted.

If you didn't count, how could you figure it out? Is there a way?

Donte: There are seven over there, so there are more here.

Why?

Donte: Because that's a larger number than this one.

Susanna: Twenty-two, because there are 22 in our class.

There are 22 kids who have zippers?

Susanna: No...

That's important information though. You're reminding us of the total number of kids in our class. How could we use that information?

Kaneisha: If we have 7 and we want 22, we put that number... Um, you can count up to that number and see how many.

Kind of like the number combinations we do.

Max: You could just count the clips. Like 7 and 15, and just count them, 1, 2, 3, all the way till you're done.

So you're trying to figure out, or check, how many people are in the whole survey.

Fernando: I counted a different way: 8 *[puts up a finger],* 9 *[adds another finger, and continues to do so with each number],* 10, 11, 12, 13, 14, 15, 16, 17. 18, 19, 20, 21, 22.

That was a lot of fingers. How did you keep track of how many to count?

Fernando: I knew I had to use all my fingers once for ten and then one hand again for five more.

Did anyone think about it a different way?

Nathan: I did it a different way. If you took 5 of the 7 ones and put it with the 15, that would make 20. Then there's 2 left from the 7, and 20 and 2 is 22.

So we discovered a lot of ways to find out that there are 22 people in this survey.

Inventing Survey Representations

What Happens

Working with another survey question, students are introduced to the idea of making representations of collected data. They first sort people, then use cubes or Kid Pins to represent the people, and finally make representations on paper with markers or stick-on materials. Students' work focuses on:

- inventing and constructing data representations
- explaining and interpreting results of surveys
- presenting data to others
- making sense out of other students' data representations

Materials

- Survey boards
- Kid Pins
- Chart paper
- Class lists (1 per student)
- Interlocking cubes (available)
- Drawing paper, 11 by 17 inches (2 sheets per student)
- Colored pencils or markers (to share)
- Construction paper, stickers or small stick-on notes, buttons, glue (optional)
- Student Sheet 7 or 8 (1 per student, homework)

Activity

Gathering Attendance Data

Gather students in a circle and ask how many are in math class today. After a few responses, count around the class to check. Determine both how many students are here today and how many are absent, as you did for the Attendance Survey (p. 122). Add today's date and numbers to your chart-paper attendance record.

	ATTENDANCE	
October 9	25 here 3 absent	28 total
October 10	26 here 2 absent	28 total

Sorting Ourselves

Now that students have had some experience with survey questions, they are going to make *representations*, or "ways of showing" their data.

Today's survey question is this: Are you wearing shoes with laces or shoes with no laces?

Let's take this survey with our bodies first, and then we will try it with Kid Pins.

Ask all students wearing shoes that have laces to stand in one corner or area of the room, and all students with no laces to stand in another. When everyone is arranged, ask how they might figure out the results of their survey.

How can we figure out how many people are in each group?

Students will probably recognize that they can find the number in each group by counting. Since people may be moving too much to count accurately, or they may be clumped in such a way that makes counting hard, ask students to suggest ways to keep track of who's been counted and who hasn't. Possible ways of keeping track include these:

- Students in each group line up and count down the line.
- Students in one group stand, then sit down one by one as they count off.
- A student who is not in the group to be counted steps forward to count. The one counting may need to point to each student in turn, or have them sit down once they've been counted.
- Students record the data on the survey boards with their Kid Pins.

Count both the "laces" and the "no laces" groups in a couple of the ways students suggest. If results differ from one count to the next, continue to count in different ways until students are confident that they know the number in each group. Record these numbers on chart paper.

When all of you got into two groups, one with laces and one without, that was one way of representing our data. We could see how many people were in each group, and we also counted them. Here's another way: Let's use our Kid Pins.

Label the survey boards with stick-on notes that read *laces* and *no laces,* with a quick sketch for visual support. If necessary, distribute the Kid Pins.

How many Kid Pins will go on the "no laces" board? How many on the "laces" board?

Many students will understand by this time that the number of pins is exactly the same as the number of people they just counted, while others will not yet realize this. In any case, they will need to check by counting.

Record the new numbers on chart paper, next to the numbers you got from directly counting the people in each group. Then summarize for students:

So far we have shown—or represented—what we found out about shoelaces in two different ways. First we showed it with our bodies: The people who had laces stood over here, and the people without laces stood over there. Next, we showed our data using Kid Pins. Both are ways of showing the very same data.

Creating Representations of Data

After you were standing in the "laces" and "no laces" groups in different parts of the classroom, when you went back to your seats, one of our representations of the shoelace data disappeared! But we still have the Kid Pins to show us how many kids are in each group.

Now, you're going to have a chance to make your own representation of the data, one you can keep yourself. Can you think of how we could show our shoelace data with paper and markers? Think of ways you could use pictures, numbers, words, or all of these.

Students might suggest drawing a shoe with laces for each person in the first group, and a shoe without laces for each person in the second group. They may want to write the number of people in each category, or make a tally mark for each person with and without laces. In the **Dialogue Box**, Surveys and Representations (p. 134), students talk about how they might use a variety of materials to represent their data.

As students are sharing their ideas, make some quick sketches on the board to demonstrate what they are suggesting, or ask students to draw their idea. Show a variety of representations, using pictures, words, and numbers to stimulate students to think about different ways that they could show the data.

If you have additional materials for making permanent representations, like construction paper, buttons, colored dot stickers, or stick-on notes, ask students how they would use these. Make them available if students would like to use them. Distribute drawing paper and markers.

For the next 20 minutes or so, students work on their individual representations of the class shoelace data. You may need to remind students that their representations should show the two different groups, and the number of people in each group. (If they are having trouble remembering the number in each group, they can refer to the survey boards or chart paper.)

Observing the Students

As they are working, observe students counting and keeping track. If they are having trouble counting, count with them. The focus of this activity is on communicating the data they collected and showing different categories of data. Look for this in their representations:

- Do students make clear visual distinctions between categories?
- Does their representation make it easy to see how many are in each category?
- Are they representing the results of this survey, or are they representing a story they've created? (Some children begin telling a story, based on pictures they've drawn, that has little to do with the data from the survey they've just participated in.)

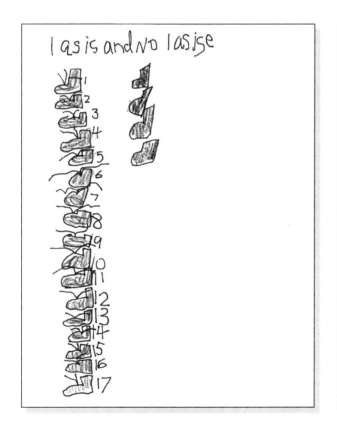

This student uses different colors as well as drawings of laces to distinguish the two categories. She has carefully counted (numbered) the shoes with laces, but feels she doesn't need to do this with the smaller number of shoes with no laces.

This student is using tally marks to show the number of children in each group. However, there's nothing to indicate what each group represents. The child knows that tallies come in groups, but is not using fives to organize the count.

This student puts the "no laces" group in a vertical stack to differentiate it from the "laces" group, which is arranged horizontally, but in a way that is hard to count.

If, after 5 or 10 minutes, some students seem to be stuck, stop the group and ask for volunteers to share what they've been doing. You may want to pair students up. Some students may prefer to make cube towers first, and use these as models for their drawing-paper representation.

When students have finished with their representations, ask them to add some writing to describe what they found out.

If there is time during Session 3 (when the work is still fresh in students' minds), gather the class together to share some of their representations.

I noticed that you found some very good ways to represent the data. Some of you used pictures, some of you used words, and some of you used numbers. Does anyone want to share what you did?

As each student shares, ask if any others recorded in the same way.

Yanni showed his solution with pictures of shoes. Did anyone else do something like that? Who did something different?

Call on a few students who have made different kinds of representations; you need not call on everyone.

Note: This is a good place to end Session 3. At the start of Session 4, once again determine the number of students in class by repeating the Attendance Survey (p. 122).

Barefoot Survey: Recording and Representing Results

Present the question for today's survey:

Did you sleep barefooted last night?

Note: If this question seems unlikely to yield mixed results in your classroom, choose another question, being sure to take into account the criteria listed in the **Teacher Note,** Data Categories (p. 125).

Distribute a one-column class list to each student for easy recording of the survey results.

We'll ask each person on the list, in order, "Did you sleep barefooted last night?" Each time we hear yes or no from someone, record that information on your sheet. It might go more quickly if you print Y for yes or N for no. Also, write a title on this sheet. That way you'll know what it's about when you look back at it later.

Go through the list slowly. Students seem to enjoy answering individually. (If anyone can't remember, ask for a best guess.) When the list is complete, students may be eager to count up the results. Encourage this curiosity, but do not rush to resolve exactly how many are in each group at this time.

Show students the materials that are available for making representations (markers and paper, as well as cubes, buttons and glue, stickers or stick-on notes, or other materials you may have). They will be working in pairs.

Work with your partner for the rest of this session to make a representation of these data we just collected. Make one that we can put on the wall, to show our survey results to other people. What are some ways you could show our results?

Students share ideas. If they get stuck, ask them to think about using pictures, numbers, or words. Some may suggest counting cubes (or other counters) into piles, lines, or stacks. Ask that anyone who uses cubes or other counters also create a representation on paper.

No matter how you decide to represent the data, everyone must show two things: (1) what the categories are—slept barefooted or not—and (2) how many people are in each category.

Some students may decide that they also want to show the name of each person. Some may write numbers to show how many in each group or the total number surveyed, but numbers are not required if the representation is visually clear.

As students work in pairs on their representations, observe as you did previously (see Observing the Students, p. 130).

If some students finish early, suggest ways they might clarify their representation even further; for example, add a few sentences about the data they have collected; add numbers; make clearer divisions between categories, and so forth.

When the representations are finished, help students display them in the room.

Sessions 3 and 4 Follow-Up

How Many of Each? As students work with totaling the numbers of data in two categories, they could continue to work on How Many of Each? problems. Student Sheets 7 and 8 offer two possibilities.

 **Homework**

On Student Sheet 7, Oranges and Cherries, students work with a total of 13. As before, this total can be modified to suit the individual needs.

On Student Sheet 8, Oranges, Cherries, and Grapes, students work with a total of three different things. Reserve this for students ready for a more challenging problem.

Sharing Results of Data Surveys This is a good opportunity to have parents and other family members see the work their children are doing with data. Invite them in to hear about what the class has been doing in this investigation. The "authors" of the various representations can be stationed by their work to explain it to their families.

 **Extension**

Surveys and Representations

These students are talking about two important concepts in this data investigation: what surveys are, and specific ways to represent data.

Yesterday we started talking about surveys. What is a survey?

Libby: When you figure out who has things and who likes things.

Good start. Who can add to that?

Garrett: It's like, you ask them. It's not just counting people. Like breakfast.

So you're remembering a survey we did earlier this year, to find out about nutrition at breakfast. Anyone else have an idea what surveys are or what they're for?

Tuan: A survey is something that you... You find information about something.

Shavonne: Yeah, like if you're trying to figure out different countries and stuff.

The other day, we were able to find out information about the number of kids in our class whose families are from different countries. So a survey is when you collect information about people. We just finished a survey about who is wearing shoelaces and who is not. What are some ways we could show what we found out?

Jonah: Draw some shoes with laces and some without laces.

How many would you draw?

Jonah: I'd draw 20 with laces and 7 with no laces.

Great. What about another way?

Shavonne: Draw cubes.

How?

Shavonne: I'd pick blue for laces and black for no laces. Then I'd just draw squares.

OK. What if you wanted to use strips of construction paper to represent this data?

Tony: I'd just take it and cut it into squares.

So how would you organize them? How many would you need?

Tony: Well, some would be laces and some would be no laces. I'd just count how many there were.

Eva: I cut out paper and I wrote on it.

What did you write?

Eva: I wrote about, like what it means. Like the brown is for laces.

Chris: I would put words on mine too. Like the name of each kid on a sticky note.

How We Got to School Today

What Happens

As students consider how they (as well as children around the world) get to school, they work with data that involve more than two categories. They brainstorm ways of representing their class data and work individually or in pairs to make their representations. Their work focuses on:

- determining how to categorize data
- making sense of data involving more than two categories
- inventing and constructing data representations
- explaining and interpreting results of surveys

Materials

- *This Is the Way We Go to School* (optional)
- Drawing materials
- Kid Pins and survey boards
- Student Sheet 9 (1 per student)
- Class lists (1 per student)

Note: For the How We Got to School survey, the class will need to collect data that falls into more than two categories. If you know that all students get to school the same way (for example, all by bus, or all by car), think of some ways to break these categories down. For example, children who came by car may have come in a truck or a van. If they all come by bus, maybe they come on different buses. Aim for at least three and no more than five categories.

Announce that today, the class will be doing a slightly different kind of survey: they'll be asking a question with more than two possible answers.

Today's survey is about how children in this class got to school today. Let's think about the different ways children get from home to school.

If you have the book *This Is the Way We Go to School*, read it to the class. This picture book illustrates a wide variety of interesting ways that children around the world travel to school. If the book is unavailable, describe a few different modes of transportation, including both common and unusual types.

Some ways of getting to school may be a surprise to your students. In Venice, Italy, the streets are canals, so children may get to school by vaporetto (a large motorboat). Some children in Norway may ski to school. In isolated areas of northern Canada, they may ride a snowmobile. Ask students if they can name any other ways that children get to school.

If you have the book, ask students about the similarities and differences between their own ways of getting to school and ways that children in other countries get to school. What do they think is the most unusual way of getting to school?

We're going to do a survey about how we got to school today. What do you think the ways will be for our class?

As children contribute suggestions, write down the main categories on the board (for example: car, bus, walk, bike), each with a sketch to help beginning readers remember what these words mean.

Activity

Collecting the Data

Distribute a copy of your class list to each student. Ask them to put the survey title, How We Got to School, at the top.

Now we are going to collect data about how we got to school today. We'll use the class list, just the way we did for our last survey. Decide which of these categories you belong in *[point to the choices on the board]*, **based on how you got to school this morning.**

Data collection goes more quickly if children write an abbreviation (such as C for car, B for bus) next to each person's name. Students will enjoy this, as long as the pace is slow enough that they can keep up.

As before, students may be eager to count up the results right after the survey. Encourage them to do so, but don't push for class agreement on the numbers for each group, because their individual work counting and combining can provide you with some assessment information.

Note: If students want to use the Kid Pins for this survey, you will need additional survey boards, one for each category. Alternatively, you might divide the survey boards into smaller areas for each category.

Refer students to their to posted representations from the earlier survey.

What are some ways you used paper last time to show the results of the survey about sleeping barefoot? Has anyone thought of some new ways? Today you're going to start new representations, to show how we got to school.

How would you use stick-on notes, or maybe construction paper?

Distribute paper and markers and make available other materials that students request. Students work individually or in pairs to make their representations. Some students might want to draw one vehicle per student, or draw one vehicle and write the number of students who got to school that way. Others may want to draw stacks of cubes for each data category. The **Dialogue Box**, What Stands for What? (p. 142) demonstrates a student in the process of using first cubes, then pencil and paper, to show the results of her data.

Teacher Checkpoint

Representations of How We Got to School

Observing the Students

As students are working, and later when you look at students' complete representations, look for the following:

- Does their representation clearly show what the survey is about?
- Does it clearly delineate categories? Is it easy to tell which pieces of data belong in each category?
- Does the representation show how many are in each category? Do students use pictures, numbers, words, or a combination?

Allow about 20-30 minutes for students to complete their representations. As they finish, ask them to write a title and a sentence that describes what they found out. You may also ask students to show: Which ways have the most and the least? How many have answered the survey? See the **Teacher Note,** Student Representations of Getting to School (p. 140), for examples of student work.

As students finish their representations, distribute Student Sheet 9, How We Got to School. Ask students to complete the chart, using only their representation for reference, not their class list. This will help them see whether they have put enough information into their representations.

❖ **Tip for the Linguistically Diverse Classroom** Pair second-language learners with students who have a clear sense of how to approach this task.

Activity

Discussing the Representations

At this point, gather students together as a full class to share their representations. Ask some volunteers to share their work. The **Dialogue Box,** Sharing Representations (p. 143), illustrates how one teacher led this discussion.

Tell us what you did and what your representation shows.

This is a good time to provide feedback to students. You may want to comment on aspects of the student's work that are particularly clear or unique, and point out aspects that need further clarification. It is not necessary to hear from every student, but encourage students to look for similarities in their work and that of others. Ask students to hold up their representations so that others may see the similar features.

Did anyone else make rows of buses and cars, the way Kristi Ann did?

You may want to allow students additional time to add any finishing touches that make sense to them, as a result of the discussion.

Choosing Student Work to Save

As the unit ends, you may want to use one of the following options for creating a record of students' work on this unit.

■ Students look back through their folders or notebooks and think about what they learned in this unit, what they remember most, and what was hard or easy for them. You might have students discuss this with partners or share in the whole group.

■ Depending on how you organize and collect student work, students might select some examples of their work to keep in a math portfolio. In addition you could make your own selection of representative work from each student's folder. Their work on Eleven Fruits (p. 92), How Many in All? (p. 106), the patterns they made for the pattern exhibit (p. 77), and Representations of How We Got to School (p. 137), can be useful pieces for assessing student growth over the school year. You may want to keep the original and make copies of these pieces for students to take home (or vice versa).

■ Send a selection of work home for families to see. Some teachers include a short letter, summarizing the work in this unit. You could enlist the help of your students and together generate a letter that describes the mathematics they were involved in. This work should be returned if you are keeping a year-long portfolio of mathematics work for each student.

Note: As you finish this unit, consider saving the Kid Pins and survey boards, either for ongoing Classroom Routines in exploring data or for use in the grade 1 unit on Collecting and Sorting Data, *Survey Questions and Secret Rules.*

Student Representations of Getting to School

Claire's work

When first graders begin representing data in their own way, they often create unique and effective ways of communicating the results of a survey. Their work will most likely include some combination of pictures, lists, numbers, and possibly charts. Students' picture-graphs will not reflect typical conventions (such as a bar graph), nor should they be expected to.

The most important thing for students at this age is to have experiences creating representations that communicate information to someone else. Keeping a sense of audience in mind is important as they show the categories of data and draw or describe the objects within each category.

Students at this age are often interested in individual pieces of data. They may want to make their own piece of data stand out, or they may want to preserve the names of each child in their representation. Encourage them to communicate the data as clearly as possible, so somebody else can tell what they are showing.

Students in one first grade classroom got these results in their class one day: 12 kids came by car, 8 came by walking, 1 came on roller blades, 1 came by bus, and 0 came by bike.

Claire (above) has used pictures to represent four categories of data. She has not included "no bikes," perhaps because she felt that it was not useful information, since there are zero objects in this category. Although her pictures are not labeled or totaled, they are easily countable. She has included only six (instead of eight) walkers, reflecting an inaccuracy in keeping track or counting. Asking Claire to compare the number of walkers in her representation to the data she collected might help her see this.

Libby's work

Donte and
Yanni's work

Libby's representation clearly reveals the results of the survey. She communicates her system of abbreviations by using a key in which it's clear that B stands for bikes (and so on). She summarizes the data in two ways, writing the total next to a picture of the category, and also writing the total (in a different order) next to her abbreviations.

Donte and Yanni have created a representation by drawing a picture of the survey boards with the Kid Pins on them. (In their room, the teacher labeled different areas of the survey boards

with stick-on notes.) They have also written sentences that clearly summarize the results of their survey.

The main goal of making representations is to show clearly what you found out, so that someone else could understand it. These students have all organized their data to help others make sense of it. One student used mainly pictures, another used mainly words, and the other pair used a combination, but they are all working to clearly show the categories in their survey and the students or objects within each group.

What Stands for What?

When Kaneisha's class collects data about the ways they got to school that day, they find that 17 students came by car (C), 5 walked (W), and 1 took the bus. Kaneisha is using cube towers to make a representation. She has a stack of cubes in different colors, not arranged in any particular pattern. She is drawing cube outlines on her paper, but does not seem to be copying directly from the stack she has built. She counts that she already has 15 cubes, and then draws 2 more.

What do you have so far?

Kaneisha: So far I have 17.

What are they for?

Kaneisha: Of the C's. Now I have to do the W's. How many W's are there? *[She counts five W's on her class list.]* Five. So I'll use five more of these. *[She makes a stack of five cubes.]*

What does that show?

Kaneisha: How many C's, there are 17, and W's.

What does that mean about how the kids in our class get to school?

Kaneisha: I don't know.

What does C mean?

Kaneisha: Car, so how many get to school by car and it's 17. So this [stack] is how the 17 looks.

And what does this [stack of five] mean?

Kaneisha: Five people got to school by walk.

Are those all the ways that kids got to school?

Kaneisha: Yeah *[looking at her class list]*. No! I forgot Tuan. He's the only one who came by bus.

Before leaving, the teacher asks if Kaneisha could find a way to show on her paper some of the information she has been talking about. Kaneisha writes a C above the tall stack and a numeral 5 above the short one.

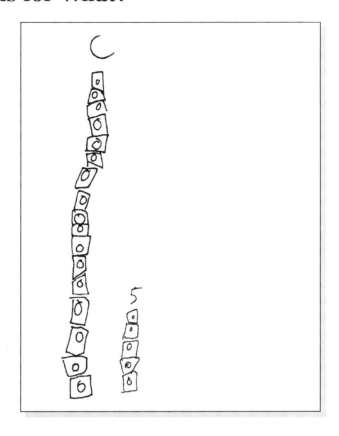

During the interaction with Kaneisha, the teacher asks about how she is using the materials and her drawings to stand for the data. Although Kaneisha is momentarily unsure when asked what her data mean, she is able to articulate her reasons for drawing particular numbers of cubes and is clear about what stands for what. Her work is accurate, though not yet complete, as she still needs to represent the student who came by bus. The process of choosing something to stand for something else is an important component of Kaneisha's data representation.

Sharing Representations

These students are sharing their representations for the final activity in Investigation 5, the How We Got to School survey.

Who would like to come up and share their work?

Mia volunteers. Her drawing (below, left) shows different modes of travel grouped together. Each group is labeled to show the number of people and how they got to school.

Mia: This shows that there are 8 people who walked, 12 people who came by car, 1 person who came by bus, and 1 person who came roller blading, and this [bottom right] shows no one came on a bike.

That's very clear. We can tell the ways that kids came to school and how many came each way. Did anyone else draw pictures of the different ways you came to school?

[Three students raise their hands.] **Can you hold up your representations?**

I see that you found different ways to show all the different ways kids came to school. Some of you used groups of pictures and some of you added words to help people understand what we found out. Who has a different way?

William comes up and shows his work (below, right) without saying anything. He has pictures scattered around the page.

What does your representation tell us about the ways that kids in our class got to school today?

William: The kids who came by car and walking and bus.

How many kids came by car?

William counts his cars. Because it is difficult to keep track, he gets 11 cars.

It's a little hard to count those carefully. It would be helpful if you could see which ones go together.

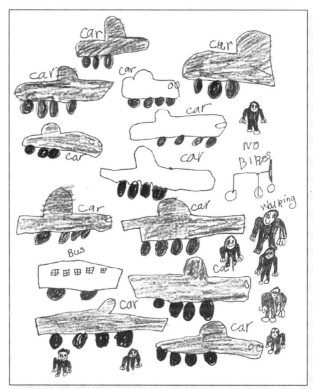

⬚D⬚I⬚A⬚L⬚O⬚G⬚U⬚E⬚ ⬚B⬚O⬚X⬚

What are some ways to show which pictures go in each group?

Chanthou shows her paper, which is organized by neatly drawn lines, putting each category in a separate box. She has included a title, "The Ways We Get to School."

Chanthou: I divided it up by lines to show each group. Then I wrote the name of the group, like bus, car.

So that's one idea for making clear what the groups are.

Did anyone else use words in their representation?

Yukiko: I wrote in Japanese and then Michelle wrote in English, so that everyone in our class could understand what it means.

We've seen a lot of different ways to represent the ways that our class gets to school. Some of you used pictures, some used words, some used numbers, and some of you combined these. You are inventing ways that are very clear.

If you would like to add anything to your representation, you may do that now. Then we'll hang them up.

Counting

Counting is an important focus in the grade 1 *Investigations* curriculum, as it provides the basis for much of mathematical understanding. As students count, they are learning how our number system is constructed, and they are building the knowledge they need to begin to solve numerical problems. They are also developing critical understandings about how numbers are related to each other and how the counting sequence is related to the quantities they are counting.

Counting routines can be used to support and extend the counting work that students do in the *Investigations* curriculum. As students work with counting routines, they gain regular practice with counting in familiar classroom contexts, as they use counting to describe the quantities in their environment and to solve problems based on situations that arise throughout the school day.

How Many Are Here Today?

Since you must take attendance every day, this is a good time to look at the number of students in the classroom in a variety of ways.

Ask students to look around and make an estimate of how many are here today. Then ask them to count.

At the beginning of the year, students will probably find the number at school today by counting each student present. To help them think about ways to count accurately, you can ask questions like these:

How do we know we counted accurately? What are different methods we could use to keep track and make sure we have an exact count? (For example, you could count around a circle of seated students, with each student in turn saying the next number. Or, all students could start by standing up, then sit down in turn as each says the next number.)

Is there another way we could count to double-check? (For example, if you counted around the circle one way, you could count around the circle the other way. If you are using the standing up/sitting down method, you could recount in a different order.)

You might want to count at other times of the day, too, especially when several students are out of the room. For example, suppose groups of students are called to the nurse's office for hearing examinations. Each time a new group of students leaves, you might ask the class to look around and think about how many students are in the room now:

So, this time Diego's table and Mia's table both went to the nurse. Usually we have 28 students here. Look around. What do you think? Don't count. Just tell me about how many students might be here now. Do you think there are more than 5? more than 10? more than 20?

Later in the year, some students may be able to use some of the information they know about the total number of students in the class and how many students are absent to reason about the number present. For example, suppose 26 students are in class on Monday, with 2 students absent. On Tuesday, one of those students comes back to school. How many students are in class today? Some students may still not be sure without counting from one, but other students may be able to reason by counting on or counting back, comparing yesterday and today. For example, a student might solve the problem in this way:

> "Yesterday we had 26 students, and Michelle and Chris were both absent. Today, Chris came back, so we have one more person, so there must be 27 today."

Another might solve it this way:

> "Well we have 28 students in our class when everyone's here. Now only Michelle is absent, so it's one less. So it's 27."

From time to time, you might keep a chart of attendance over a week or so, as shown below. This helps students become familiar with different combinations of numbers that make the same total. If you have been doing any graphing, you might want to present the information in graph form.

Day	Date	Present	Absent	Total
Monday	March 2	26	2	28
Tuesday	March 3	27	1	28
Wednesday	March 4	27	1	28
Thursday	March 5	27	1	28
Friday	March 6	28	0	28
Monday	March 9	28	0	28
Tuesday	March 10	26	2	28
Wednesday	March 11	25	3	28

After a week or two, look back over the data you have collected. Ask questions about how things have changed over time.

In two weeks of attendance data, what changes? What stays the same?

On which day were the most students here? How can you tell? Which day shows the least students here? What part of the [chart] gives you that information?

Another idea (for work with smaller numbers) is to keep track of the number of girls and boys present and absent each day. Again, many students will count by 1's. Later in the year, some will also reason about these numbers:

> "There are two people absent today and they're both girls. We usually have 14 girls, and Kaneisha's sick, that's 13, and Claire's sick, that's 12."

Can Everyone Have a Partner?

Attendance can be an occasion for students to think about making groups of two:

We have 26 students here today. Do you think that everyone can have a partner if we have 26 students?

Students can come up with different strategies for solving this problem. They might draw 26 stick figures, then circle them in 2's. They could count out 26 cubes, then put them together in pairs. They might arrange themselves in 2's, or count by 2's .

At the beginning of the year, many of your students will need to count by 1's from the beginning each time you add two more students, but gradually some will begin to notice which numbers can be broken up into pairs:

> "I know 13 doesn't work, because you can do it with 12, and 13's one more, so you can't do it."

Some students will begin to count by 2's, at least for the beginning of the counting sequence. Then, as the numbers get higher, they may still be able to keep track of the 2's, but need to count by 1's:

> "So, that's 2, 4, 6, 8, 10, 12, um, 13, 14 . . . 15, 16."

As you explore 2's with your students, keep in mind that many of them will need to return to 1's as a way to be sure. Even though some students learn the counting sequence 2, 4, 6, 8, 10, 12 . . . by rote, they may not connect this counting sequence to the quantities it represents at each step.

One teacher found a way to help students develop meaning for counting by 2's. She took photographs of each student, backed them with cardboard, then used them during the morning meeting as a model for making pairs. She laid out the photos in two columns, and asked about the new total after the addition of each pair:

We have 10 photos out so far. The next two photos are for William and Yanni. When we put those two photos down, how many photos will we have?

Lining up is another time to explore making pairs. Before lining up, count how many students are in class (especially if it's different from when you took attendance). Ask students whether they think the class will be able to line up in even pairs. For many first grade students, the whole class is too many people to think about. You can ask about smaller groups:

What if Kristi Ann's table lines up first? Do you think we could make even partners with the people at that table?

What about Shavonne's table? … Do you think Shavonne's table will have an extra? How do you know?

Is there another table that would have an extra that we could match up with the extra person from Shavonne's table?

Once students are lined up in pairs, they can count off by 2's. Because most first graders will need to hear all the numbers to keep track of how the counting matches the number of people, ask them to say the first number in the pair softly and the second one loudly. Thus the first pair in line can say, "1, **2**," the second pair can say, "3, **4**," and so forth.

Counting to Solve Problems

Be alert to classroom activities that lend themselves to a regular focus on solving problems through counting. Use these situations as contexts for counting and keeping track, estimating small quantities, breaking quantities into parts, and solving problems by counting up or back. For example, take a daily milk count:

Everyone who is buying milk today stand up. Without counting yet, who has an idea how many students might be standing up? Is it more than 5? more than 10? more than 50? … Now, let's count. How could we keep track today so that we get an accurate count?

You can make a problem out of lunch count:

We found out that 23 students are buying school lunch today. We have 27 students here. So how many students brought their own lunch from home today?

Watch for the occasional sharing situation:

Claire brought in some cookies she made to share for snack. She brought 36 cookies. Is that enough for everyone to have one cookie, including me and our student teacher? Oh, and Claire wants to invite her little brother to snack. Do we have enough for him, too? Will there be any cookies left over?

The sharing of curriculum materials can also be the basis of a problem:

Each pair of students needs a deck of number cards to share. While I'm getting things together, work on this problem with your partner. We said this morning that we have 26 students here. If I need one deck for each pair, how many decks do I need?

Exploring Data

Through data routines at grade 1, students gain experience working with categorical data—information that falls into categories based on a common feature (for example, a color, a shape, or a shared function). The data routines specifically extend work students do in the *Investigations* curriculum. The Guess My Rule game and its many variations (introduced in the unit *Survey Questions and Secret Rules*) can be used throughout the year for practice with organizing sets into categories and finding ways to describe those categories—a fundamental part of analyzing data. Students can also practice collecting and organizing categorical data with quick class surveys that focus on their everyday experiences; this practice supports the survey-taking they do in the curriculum.

Guess My Rule

Guess My Rule is a classification game in which players try to figure out the common characteristic, or attribute, of a set of objects. To play the game, the rule maker (who may be the teacher, a student, or a small group) decides on a secret rule for classifying a particular group of things. For example, a rule for classifying people might be WEARING STRIPES.

The rule maker (always the teacher when the game is first being introduced) starts the game by giving some examples of objects or people who fit the rule. The guessers then try to find other items that fit the same rule. Each item (or person) guessed is added to one of two groups—either *does fit* or *does not fit* the rule. Both groups must remain clearly visible to the guessers so they can make use of all the evidence as they try to figure out the rule.

Emphasize to the players that "wrong" guesses are as important as "right" guesses because they provide useful clues for finding the rule. When you think most students know the rule, ask for volunteers to share their ideas with the class.

Once your class is comfortable with the activity, students can choose the rules. Initially, you may need to help students choose appropriate rules.

Guess my Rule with People When sorting people according to a secret rule, always base the rule on just one feature that is clearly visible, such as WEARING A SHIRT WITH BUTTONS, or WEARING BLUE. When students are choosing the rule, they may choose rules that are too obvious (such as BOY/GIRL), so vague as to apply to nearly everyone (WEARING DIFFERENT COLORS), or too obscure (HAS AN UNTIED SHOELACE). Guide and support students in choosing rules that work.

Guess My Rule with Objects Class sets of attribute blocks (blocks with particular variations in size, shape, color, and thickness) are a natural choice for Guess My Rule. You can also use collections of objects, such as sets of keys, household container lids, or buttons. One student sorts four to eight objects according to a secret rule. Others take turns choosing an object from the collection that they think fits the rule and placing it in the appropriate group. If the object does not fit, the rule maker moves it to the NOT group. After several objects have been correctly placed, students can begin guessing the rule.

Guess My Object Once students are familiar with Guess My Rule, they can use the categories they have been identifying to play another guessing game that also involves thinking about attributes. In this routine, students guess, by the process of elimination, which particular one of a set of objects has been secretly chosen. This works well with attribute blocks or object collections.

To start, place about 20 objects where everyone can see them. The chooser secretly selects one of the objects on display, but does not tell which one (you may want the chooser to tell you, privately). Other students ask yes-or-no questions, based on attributes, to get clues to help them identify the chosen object. After each answer, students move to one side the objects that have been eliminated. That is, if someone asks "Is it round?" and the answer is yes, all objects that are *not* round are moved aside.

Pause periodically to discuss which questions help eliminate the most objects. For example, "Is it this one?" eliminates only one object, whereas "Is it red?" may eliminate several objects. For more challenge, students can play with the goal of identifying the secret object with the fewest questions.

Quick Surveys

Class surveys can be particularly engaging when they connect to activities that arise as a regular part of the school day, and they can be used to help with class decisions. As students take surveys and analyze the results, they get good practice with collecting, representing, and interpreting categorical data.

Early in first grade, to keep the surveys quick and the routine short, use questions that have exactly two possible responses. For example:

Would you rather go outside or stay inside for recess today?

Will you drink milk with your lunch today?

Do you need left-handed or right-handed scissors?

As the school year progresses, you might include some survey questions that are likely to have more than two responses:

Which of these three books do you want me to read for story time?

Who was your teacher last year?

Which is your favorite vegetable growing in our class garden?

How old are you?

In which season were you born?

Try to choose questions with a predictable list of just a few responses. A question like "What is your favorite ice cream flavor?" may bring up such a wide range of responses that the resulting data is hard to organize and analyze.

As students become more familiar with classroom surveys, invite the class to brainstorm questions with you. You may decide to avoid survey questions about sensitive issues such as families, the body, or abilities, or you might decide to use surveys as a way of carefully raising some of these issues. In either case, it is best to avoid questions about material possessions ("Does your family have a car?").

Once the question is chosen, decide how to collect and represent data. Be sure to vary the approach. One time, you might collect data by recording students' responses on a class list. Another time, you might take a red interlocking cube for each student who makes one response, a blue cube for each student who makes the other response. Another time, you might draw pictures. If you have prepared Kid Pins and survey boards for use in *Mathematical Thinking at Grade 1,* these can be used for collecting the data from quick surveys all year.

Initially, you may need to help students organize the collected data, perhaps by stacking cubes into "bars" for a "graph," or by making a tally. Over time, students can take on more responsibility for collecting and organizing the data.

Always spend a little time asking students to describe, compare, and interpret the data.

What do you notice about these data?

Which group has the most? the least? How many more students want [recess indoors today]?

Why do you suppose more would rather [stay inside]? Do you think we'd get similar data if we collected on a different day? What if we did the same survey in another class?

Understanding Time and Changes

These routines help students develop an understanding of time-related ideas such as sequencing of events, understanding relationships among time periods, and identifying important times in their day.

Young students' understanding of time is often limited to their own direct experiences with how important events in time are related to each other. For example, explaining that an event will occur *after* a child's birthday or *before* an important holiday will help place that event in time for a child. Similarly, on a daily basis, it helps to relate an event to a benchmark time, such as *before* or *after* lunch. Both calendars and daily schedules are useful tools in sequencing events over time and preparing students for upcoming events. These routines help young students gain a sense of basic units of time and the passage of time.

Calendar

The calendar is a real-world tool that people use to keep track of time. As students work with the calendar, they become more familiar with the sequence of days, weeks, and months, and the relationships among these periods of time. Calendar activities can also help students become more familiar with relationships among the numbers 1–31.

Exploring the Monthly Calendar At the start of each month, post the monthly calendar and ask students what they notice about it. Some students might focus on arrangement of numbers or total number of days, while others might note special events marked on the calendar, or pictures or designs on the calendar. All these kinds of observations help students become familiar with time and ways that we keep track of time. You might record students' observations and post them near the calendar.

As the year progresses, encourage students to make comparisons between the months. Post the calendar for the new month next to the calendar for the month just ending and ask students to share their ideas about how the two calendars are similar and different.

Months and Years To help students see that months are part of a larger whole, display the entire calendar year on a large sheet of paper. Cut a small calendar into individual monthly pages and post the sequence of months on the wall. You might decide to post the months according to the school year, September through August, or the calendar year January through December. At the start of each month, ask students to find the position of the new month on the larger display. From time to time, you might also use this display to point out dates and distances between them as you discuss future events or as you discuss time periods that span a month or more. (Last week was February vacation. How many weeks until the next vacation?)

How Many More Days? Ask students to figure out how long until special events, such as birthdays, vacations, class trips, holidays, or future dates later in the month. For example:

Today is October 5. How many more days until October 15?

How many more days until [Nathan's] birthday?

How many more days until the end of the month?

Ask students to share their strategies for finding the number of days. Initially, many students will count each subsequent day. Later, some students may begin to find their answers by using their growing knowledge of calendar structure and number relationships:

> "I knew there were three more days in this row and I added them to the three days in the next row. That's 6 more days."

Others may begin using familiar numbers such as 5 or 10 in their counting:

> "Today is the 5th. Five more days is 10, and five more is 15. That's 10 more days until October 15."

For more challenge, ask for predictions that span two calendar months. For example, you might post the calendar for next month along side of the calendar of this month and ask a question like this:

It's April 29 today. How many more days until our class trip on May 6?

Note that we can refer to a date either as October 15 or as the 15th day of October. Vary the way you refer to dates so that students become comfortable with both forms. Saying "the 15th day of October" reinforces the idea that the calendar is a way to keep track of days in a month.

How Many Days Have Passed? Ask questions that focus on events that have already occurred:

How many days have passed since [a special event]? since the weekend? since vacation?

Mixed-Up Dates If your monthly display calendar has date cards that can be removed or rearranged, choose two or three dates and change their position on the calendar so that the numbers are out of order. Ask students to fix the calendar by pointing out which dates are out of order.

Groups of two or three can play this game with each other during free time. Students can also remove all the date cards, mix them up, and reassemble the calendar in the correct order. You might mark the space for the first day of the month so that students know where to begin.

Daily Schedule

The daily schedule narrows the focus of time to hours and shows students the order of familiar events over time. Working with schedules can be challenging for many first graders, but regular opportunities to think and talk about the idea will help them begin predicting what comes next in the schedule. They will also start to see relationships between particular events in the schedule and the day as a whole.

The School Day Post a schedule for each school day. Identify important events (start of school, math, music, recess, reading, lunch) using pictures or symbols and times. Include both analog (clock face) and digital (10:15) representations. Discuss the daily schedule each day with students using words such as *before* math, *after* recess, *during* the morning, *at the end of* the school day. Later in the school year you can begin to identify the times that events occur as a way of bridging the general idea of sequential events and the actual time of day.

The Weekend Day Students can create a daily schedule, similar to the class schedule, for their weekend days. Initially they might make a "timeline" of their day, putting events in sequential order. Later in the year they might make another schedule where they indicate the approximate time of day that events occur.

Weather

Keeping track of the weather engages young students in a real-life data collection experience in which the data they collect changes over time. By displaying this ongoing collection of data in one growing representation, students can compare changes in weather across days, weeks, and months, and observe trends in weather patterns, many of which correspond to the seasons of the year.

Monthly Weather Data With the students, choose a number of weather categories (which will depend on your climate); they might include sunny, cloudy, partly cloudy, rainy, windy, and snowy.

If you vary the type of representation you use to collect monthly data, students get a chance to see how similar information can be communicated in different ways. On the following page you'll see some ways of representing data that first grade teachers have used.

At the end of each month (and periodically throughout the month), ask questions to help students analyze the data they are collecting.

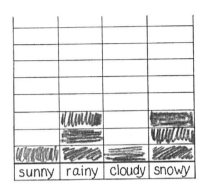

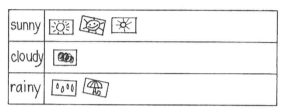

Another approach over the entire year is to pre-pare 10-by-10 grids from 1-inch graph paper, making one grid for each weather category your class has chosen. Post the grids, labeled with the identifying weather word. Each day, a student records the weather by marking off one square on one or more grids; that is, on a sunny day, the student marks a square on the "sunny" grid, and if it's also windy, he or she marks the "windy" grid, too.

Weather data can be collected on displays like these. In the second example, a student draws each day's weather on an index card to add to the graph. The third example uses stick-on dots.

What is this graph about?

What does this graph tell us about the weather this month (so far)?

What type of weather did we have for the most days? What type of weather did we hardly ever have?

How is the weather this month different from the weather last month? What are you looking at on the graph to help you figure that out?

How do you think the weather graph for next month will look?

Yearly Weather Data If you collect and analyze weather data for some period of time, consider extending this over the entire school year. Save your monthly weather graphs, and periodically look back to see and discuss the changes over longer periods of time.

From time to time, students can calculate the total number of days in a certain category by counting the squares. Because these are arranged in a 10-by-10 grid, some students may use the rows of 10 to help them calculate the total number of days. ("That's 10, and another 10 is 20, and 21, 22, 23.")

Making Weather Representations After students have had some experience collecting and recording data in the grade 1 curriculum (especially in *Survey Questions and Secret Rules*), they can make their own representation of the weather data. For one month, record the weather data on a piece of chart paper (or directly on your monthly calendar), without organizing it by category. At the end of the month, ask students to total the number of sunny days, rainy days, and so forth, and post this information (perhaps as a tally). Students then make their own representation of the data, using pictures, numbers, words, or a combination of these. Encourage them to use clear categories and show the number of days in each.

The following activities will help ensure that this unit is comprehensible to students who are acquiring English as a second language. The suggested approach is based on *The Natural Approach: Language Acquisition in the Classroom* by Stephen D. Krashen and Tracy D. Terrell (Alemany Press, 1983). The intent is for second-language learners to acquire new vocabulary in an active, meaningful context.

Note that *acquiring* a word is different from *learning* a word. Depending on their level of proficiency, students may be able to comprehend a word upon hearing it during an investigation, without being able to say it. Other students may be able to use the word orally, but not read or write it. The goal is to help students naturally acquire targeted vocabulary at their present level of proficiency.

We suggest using these activities just before the related investigations. The activities can also be led by English-proficient students.

For All Investigations

the numbers 1–15, color names (red, blue, yellow, green, orange), total

Adapt this activity to work on different color names, as needed. In this unit, students often identify their interlocking cubes and the pattern blocks by color. Check for the colors of the materials available to your students.

1. Show and read aloud the numbers 1–3, each written on a *red* square of paper. Identify this color.

2. Show and read aloud the numbers 4–6, each written on a *blue* square of paper. Identify this color.

3. Show and read aloud the numbers 7–9, each written on a *yellow* square of paper. Identify this color.

4. Show and read aloud the numbers 10–12, each written on a *green* square of paper. Identify this color.

5. Show and read aloud the numbers 13–15, each written on an *orange* square of paper. Identify this color.

6. As you display the numbers written on the red and blue squares, ask questions related to the numbers or the color, requiring a one-word response. For example:

   Is the number 5 *[point]* on a red square?

   Is the number 2 *[point]* on a red square?

   [Point to the 3.] **Is this a 1 or a 3?**

   Is the number 6 *[point]* on a red or blue square?

7. Continue in this manner for each of the colored squares of numbers.

8. Ask a volunteer to choose one square from the red group and one from the blue group. Make a stack of the same number of cubes in corresponding colors. Say the number in each stack, then snap the two stacks together. Ask:

   What is the *total* number of these cubes? We can count to find out.

   Continue having students choose numbers on two squares, then use the cubes to find the total. Use only the numbers 1–6 for this part of the activity.

Investigations 2-3

numbers, words, pictures, predict

1. Draw three large circles on the board. Take three medium-size stick-on notes: on the first, write the word *flower;* on the second, write the number 3; on the third, draw three flowers. Stick each note in a different circle.

2. As you point to the three examples, explain that the first shows a *word*, the second shows a *number*, and the third shows a *picture*. Tell students that you are going to be putting other examples into these three groups: *[point]* words, numbers, and pictures.

3. Write the number 7 on another stick-on note. Ask students to decide which group it goes in; then place it inside the number circle.

4. Repeat the previous step with other examples: the word *cat*, a drawing of four faces, the number 5, a drawing of two houses, the word *bus*.

5. Hiding your work from the students, collect the notes from the board. Choose two categories and place them on the table in a row, in an alternating pattern, such as *picture, number, picture,*

number, picture, number. (Make additional notes as needed.) Cover all but the first two notes with a sheet of paper.

6. Ask students to name each category as you point to the notes, starting from the end. Before you reveal the next note in your row, ask the group to *predict*: Will it be a number or a picture? After students predict, uncover the note to check. Continue asking them to predict the next category until all are revealed.

peas, carrots, blueberries, compare, larger [or bigger], smaller

1. If possible, show students real peas, carrots, and a blueberries. Otherwise, show or draw pictures of them. Identify them, then ask questions requiring a one-word response. For example:

Which are green, peas *[point]* **or carrots** *[point]*?
What color are the carrots?
Are the blueberries green?
Are the blueberries orange?

2. Make (or draw) two groups of peas, with two peas in one group and seven peas in the other. Label each group with the numbers 2 and 7. Identify the groups by number.

This group *[point]* **has 2. This group** *[point]* **has 7. Let's** *compare* **the groups.**

Ask students to name the groups by number to answer your questions.

Which group is larger [bigger]?
Which group is smaller?

Repeat, using different groups of peas, always keeping the larger group significantly larger (for example, 2 peas and 8 peas; 3 peas and 10 peas). Each time, label the group with a number before asking students to compare.

Blackline Masters

_____, 19_____

Dear Family,

We are starting a unit called *Mathematical Thinking at Grade 1*. During this unit, your child will use lots of mathematical tools and materials, including interlocking cubes, pattern blocks, and Geoblocks, while learning ways of working mathematically.

Here are some of the things we will be doing in class:

■ In their study of number, the children will play mathematical games and solve problems in which they count, find which of two numbers is larger (8 is more than 3), and find the total of two or more numbers (2 and 6 is 8).

■ To investigate geometry concepts, the children will create designs, buildings, and patterns with the pattern blocks and Geoblocks, finding how they fit together and thinking about how the shapes are alike and different. They also try to predict what comes next in a pattern like this one:

□ △ △ □ △ △ □ △ △   What comes next?

■ An important part of mathematics is the study of data. The children will collect and organize data about themselves as a group. For example: How did we get to school today? 12 of us walked, 9 came by car, and 6 rode the bus.

While our class is working on this unit and throughout the year, you can help in several ways.

■ Children work out number problems by using real objects. At home, try to provide a collection of small objects for counting, such as beans, buttons, or pennies.

■ Your child will bring home three number games. Play the games frequently with your child. Find a safe place to store the number cards and game directions, perhaps in an empty folder or manila envelope.

■ When working on problems at home, your child may use pictures, numbers, words, or a combination of these to keep track of the work. All are important ways of showing mathematical thinking. Let your child find his or her own ways to solve problems and record work.

We are looking forward to an exciting few weeks as we create a mathematical community in our classroom.

Sincerely,

Compare

Materials: Deck of Number Cards 0–10
(remove the Wild Cards)

Players: 2

Object: Decide which of two cards shows
a larger number.

How to Play

1. Mix the cards and deal them evenly to each player. Place your stack of cards facedown in front of you.

2. At the same time, both of you turn over the top card in your stack. Look at the numbers. If your number is larger, you say "Me!" If the two cards are the same, turn over the next card.

3. Keep turning over cards. Each time, say "Me!" if your number is larger.

4. The game is over when you have both turned over all the cards in your stack.

Variations

a. If you have the **smaller** number, you say "Me."

b. Play with three people. Look at all three numbers. If you have the largest number, say "Me."

c. Add the four wild cards to the deck. A wild card can be made into any number.

Double Compare

Materials: Deck of Number Cards 0–10
                 (remove the wild cards)

Players: 2

Object: Decide which of two totals is greater.

> **Note to Families**
> In this game, your child will be finding the totals of pairs of numbers. You will need a set of Number Cards to play this game.

How to Play

1. Mix the cards and deal them evenly to each player. Place your stack of cards facedown in front of you.

2. At the same time, both of you turn over the top two cards in your stack. Look at your two numbers and find the total. Then find the total of the other player's numbers.

   If your total is more than the other player's, say "Me!" If the two totals are the same, turn over the next two cards.

3. Keep turning over two cards. Say "Me!" each time your total is more.

4. The game is over when you have both turned over all the cards in your stack.

Variations

a. If your total is **less,** say "Me."

b. Play with three people. Find all three totals. If yours is the most, say "Me."

c. Add the four wild cards to the deck. A wild card can be made into any number.

Balls and Bats

Note to Families
There are many ways to solve this problem. Encourage your child to find his or her own ways to solve the problem and record the work.

I have ___8___ balls and bats.

How many of each could I have?

Keep track of your work. You can use pictures, numbers, or words.

STAIRCASE CARDS

12	12		
11	11		
10	10		
9	9		
8	8	19 20	19 20
7	7	18 19	18 19
6	6	17 18	17 18
5	5	16 17	16 17
4	4	15 16	15 16
3	3	14 15	14 15
2	2	13 14	13 14
1	1	13	13

Patterns from Home

Over the next few days, we are going to set up a pattern exhibit in the classroom. The children have been making patterns with colored cubes.

Help your child look around for something to bring in for our pattern exhibit. Please don't send anything breakable. Your child can bring everything home after the exhibit.

Here are some places to look for patterns:
Clothing (a scarf, a t-shirt, a necktie)
Wrapping paper
Towels or napkins
Book covers

Your child could also make patterns, look for a pattern to cut out of a magazine, or make a drawing of a pattern in your home that you can't bring into school (floor tile, wallpaper).

I'm sure you'll find other things we haven't thought of.

Thanks for your help!

CUBE PATTERN STRIPS

PATTERN BLOCK CUTOUTS (page 1 of 6)

Duplicate these hexagons on yellow paper and cut apart.

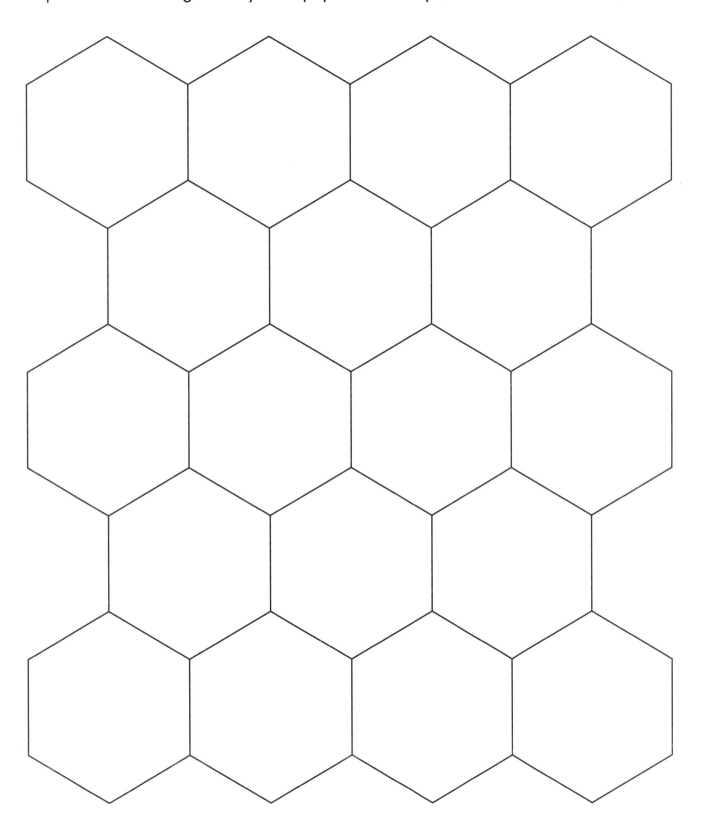

PATTERN BLOCK CUTOUTS (page 2 of 6)

Duplicate these trapezoids on red paper and cut apart.

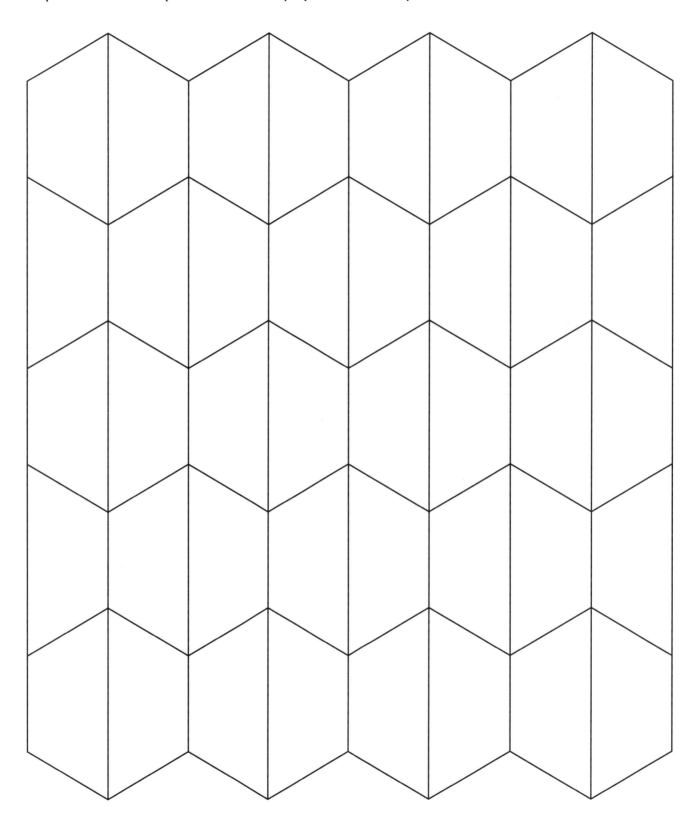

Investigation 3 Resource
Mathematical Thinking at Grade 1

Duplicate these triangles on green paper and cut apart.

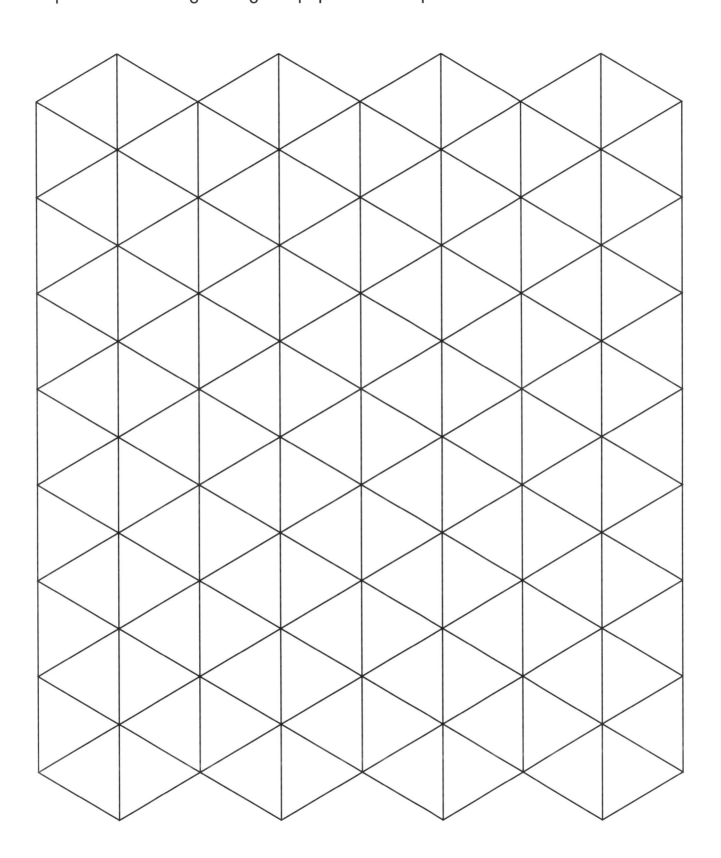

Duplicate these squares on orange paper and cut apart.

PATTERN BLOCK CUTOUTS (page 5 of 6)

Duplicate these rhombuses on blue paper and cut apart.

Investigation 3 Resource
Mathematical Thinking at Grade 1

Duplicate these rhombuses on tan paper and cut apart.

Collect 15 Together

Materials: One dot cube
            25 counters

Players: 2

Object: With a partner, collect 15 counters.

How to Play

1. To start, one player rolls the dot cube. What number did you roll? Take that many counters.

2. Take turns rolling the dot cube. Take that many counters and add them to the collection.

3. After each turn, check the total number of counters in your collection. The game ends when you have 15 counters.

Variations

a. At the end of each game, determine how many **more** than 15 counters you have.

b. Play Collect 25 Together or Collect 40 Together.

c. Collect **exactly** 15 counters. If the number you roll takes you over 15, skip that turn and roll again.

d. For each turn, write the number you rolled and the total number of counters you have so far.

e. Play with three people.

f. Instead of a dot cube, use the Number Cards for 1 to 6. Mix them and turn up one at a time.

Dinosaurs and Tigers

I have ___10___ dinosaurs and tigers.

How many of each could I have?

Keep track of your work. You can use pictures, numbers, or words.

Note to Families

There are many ways to solve this problem. Encourage your child to find his or her own ways to solve the problem and record the work.

Oranges and Cherries

I have ___13___ oranges and cherries.

How many of each could I have?

Keep track of your work. You can use pictures, numbers, or words.

Note to Families
There are many ways to solve this problem. Encourage your child to find his or her own ways to solve the problem and record the work.

Oranges, Cherries, and Grapes

I have __9__ oranges, cherries, and grapes.

How many of each could I have?

Keep track of your work. You can use pictures, numbers, or words.

Note to Families
There are many ways to solve this problem. Encourage your child to find his or her own ways to solve the problem and record the work.

Name _____   Date _____

How We Got to School

Ways we got to school today   How many children?

_____   _____

_____   _____

_____   _____

_____   _____

_____   _____

How many children were in your survey?

_____

What was the way most children got to school?

_____

What was the way the fewest children got to school?

_____

0	0	0	0
1	1	1	1
2	2	2	2

3	3	3	3
4	4	4	4
5	5	5	5

6	6	6	6
7	7	7	7
8	8	8	8

9	9	9	9

10	10	10	10

Wild Card	Wild Card	Wild Card	Wild Card

Game Record Sheet

Game: _____

_____

Play this game at home. You should have the directions and other things you need. After you play, fill out and return this sheet.

Note to Families
Please play this game with your child, then help fill out and return this sheet. The more times children play a mathematical game, the more practice they get with important skills and with reasoning mathematically.

Who played the game?

_____

Write about what happened when you played the game.

_____

_____

_____

_____

_____

_____

Practice Pages

This section provides optional homework for teachers who want or need to give more homework than is suggested to accompany the activities in this unit. With the games or problems included here, students get additional practice in learning about number relationships and solving number problems. Whether or not the *Investigations* unit you are presenting in class focuses on number skills, continued work at home on developing number sense will benefit students. In this unit, practice pages include the following:

How Many of Each? Problems This type of problem is introduced in Investigation 2 of this unit. Five additional problems are provided here. (None of these pages should be sent home until students have had some experience in class with this type of problem.) You can modify any of the numbers and make up new problems in this format, using numbers that are appropriate for your students. Students need not always work with a different total each time they do a How Many of Each? problem. Repeating the same total with different objects gives them more practice with number combinations.

Practice Page A

I have ___9___ red and blue crayons.

How many of each could I have?

Keep track of your work. You can use pictures, numbers, or words.

Practice Page B

I have ___10___ bananas and strawberries.

How many of each could I have?

Keep track of your work. You can use
pictures, numbers, or words.

Practice Page C

I have ___12___ cars and boats.

How many of each could I have?

Keep track of your work. You can use pictures, numbers, or words.

Practice Page D

I have ___12___ cows and pigs.

How many of each could I have?

Keep track of your work. You can use pictures, numbers, or words.

Practice Page E

I have ___14___ peas and carrots.

How many of each could I have?

Keep track of your work. You can use pictures, numbers, or words.